The 5S Innovation Model

This book is aimed to help us look into the future of mining by defining ultimate operational conditions that will be present in a typical mining operation regardless of how far in the future. It introduces an innovation strategy designed to identify current and future technologies to achieve specific ultimate operational conditions that will be present in 'the mine of the future'.

The mining innovation strategy presented here is the result of several innovation projects where the author had the opportunity to assist and had successfully implemented it at several mining companies and mining research institutions around the world, including Australia, South Africa, the United States, Canada, Peru, and Mexico. This innovation strategy is designed to be consistent with any type of mining method as well as any commodity being mined, such as metal or nonmetal mining, soft-rock or hard-rock mining, underground or surface mining.

The five ultimate mining operational conditions or drivers discussed in this book were carefully defined considering current operational and technology trends, which will keep any mining company competitive during the following decades. The mining innovation strategy thus considers five ultimate operational conditions or drivers (1) Achieving maximum safety, (2) simplifying systems, (3) using smart-intelligent systems, (4) designing stealth operations, and (5) sustainable use of environmental and human resources within the operation. These five innovation drivers are common denominators to any mining method, regardless of their operational nature or commodity being mined either today or in the future.

It is thus envisaged that the mining innovation model introduced in this book can serve as an initial guideline for the mining industry to better identify current and future technologies that need to be addressed today.

The 5S Innovation Model

A Tech-Innovation Strategy for the Mine of the Future

Antonio 'Tony' Nieto

CRC Press
Taylor & Francis Group
Boca Raton London New York

CRC Press is an imprint of the
Taylor & Francis Group, an **informa** business

A BALKEMA BOOK

First published 2024
by CRC Press/Balkema
4 Park Square, Milton Park, Abingdon, Oxon, OX14 4RN

and by CRC Press/Balkema
2385 NW Executive Center Drive, Suite 320, Boca Raton FL 33431

CRC Press/Balkema is an imprint of the Taylor & Francis Group, an informa business

British Library Cataloguing-in-Publication Data
A catalogue record for this book is available from the British Library

ISBN: 978-1-032-62267-5 (hbk)
ISBN: 978-1-032-62268-2 (pbk)
ISBN: 978-1-032-62269-9 (ebk)

DOI: 10.1201/9781032622699

Typeset in Times New Roman
by Apex CoVantage, LLC

Contents

Preface

Mining and human development

The first human activities crucial for survival were most likely farming, fishing, animal domestication, and mining. For most of human history, our ancestors survived by hunting animals for food and gathering plants, wood, minerals, and other resources from the environment. As humans learned to cultivate crops, breed animals for food and labor, and fish for sustenance, these activities allowed humans to settle in one place, establish communities, and develop civilizations.

Mining and manufacturing also emerged, as humans developed tools and technology to extract and process raw materials, such as metals and minerals, and to create a wide range of utensils and products for hunting and labor. Mining has been an essential aspect of human development for thousands of years and continues to play a critical role in modern society. Mining provides essential raw materials that are used in a wide range of products and technologies, including construction materials, electronics, vehicles, medical equipment, and energy production.

Mining is a key driver of technological development; the discovery and extraction of new minerals and metals have played a key role in technological progress throughout history. For example, the discovery of copper and tin, and later iron, allowed humans to develop more advanced tools and weapons, whereas modern mining techniques to economically extract and refine copper, lithium, cobalt, zinc, nickel, and other key metals have enabled the production of advanced technologies such as smartphones and electric cars.

The mining industry creates jobs and generates economic growth, particularly in the areas where other industries may be limited. This can help support local communities and contribute to the overall development of regions and countries. Mining has helped connect different regions of the world through exchange of raw materials and finished products, leading to increased international trade and promoting economic development on a global scale.

Mining is a complex and multi-faceted industry that involves the extraction, processing, and distribution of minerals and metals. It encompasses a wide range of activities, from exploration and prospecting to mine design and construction and from mining operations to environmental management and rehabilitation.

In addition to providing essential raw materials for products and technologies, mining also contributes to the global economy by generating jobs and revenue. The industry employs millions of people around the world, and the products and technologies it produces are used in virtually every sector of the economy.

However, mining can also have negative impacts on the environment and local communities, such as deforestation, water pollution, and land degradation. These impacts can be mitigated through responsible mining practices, including minimizing waste, reducing energy consumption, and restoring ecosystems after mining operations have ceased.

The mining industry is constantly evolving, driven by technological advancements, and changing societal expectations. As the demand for minerals and metals continues to grow, the industry is increasingly focused on sustainability and responsible resource management, seeking to minimize its environmental impact while maximizing its social and economic benefits.

Overall, mining is a vital industry that plays a critical role in human development and modern society, and it will continue to do so for the foreseeable future.

Acknowledgments

Dear readers,

It is with great pleasure to present my new book "Innovation Strategy for the Minerals Industry, The 5S Innovation Model". As with any creative endeavor, this book would not have been possible without the support and contributions of many people who have helped to make this project a reality. My gratitude to Greg Dewicki, Admundo LaPorte, Jesus Castillo, and Natasha Xenaki Diego Maura, for their motivation to write this book.

I want to express my sincere gratitude to my parents and to my good friends and colleagues, David Coolbaugh, Donald Gentry, Kadri Dagdelen, Margaret Armstrong, Vlad Kecojevic, Turgay Ertekin, and Jaime Lomelin for their support, and valuable insights, throughout my career. Their advice and encouragement helped shape me as a mining professional, academic, and researcher, all merging into an unexpected holistic combination of new ideas and views about mining which are reflected in this book.

I would like to thank The Guanajuato School of Mines, and The Colorado School of Mines for their academic training. I am especially grateful to Wits University for their bestowed professorship and tenure. My infinite gratitude to my wife, Karina, for her constant support and encouragement. Her patience, understanding, and unwavering belief in me helped me push through difficult moments in my life and keep my focus on believing in myself, regardless of the many obstacles I have faced.

I would also like to express my deepest gratitude to all my mining professional followers and readers, whose support and interest in my work inspire me to keep documenting and sharing my ideas with the world.

At last, I want to thank those who consistently were pouring water onto the fire and thereby steaming up my passion to continue my career.

I dedicate this book to my beloved students at Virginia Tech, Penn State, and Wits University. Thank you all for being a part of my journey.

Sincerely,
Prof. Dr. Antonio 'Tony' Nieto

Author biography

Prof. Dr. Antonio 'Tony' Nieto is a highly experienced mining engineer who has dedicated his career to advancing the mining industry. With over 30 years of experience, he has played a key role in numerous mining projects across the Americas, Europe, and Africa. Dr. Nieto's area of expertise lies in designing and implementing projects that enhance the safety and efficiency of mining operations, as well as developing innovative technologies for this purpose. He is a leading figure in the mining industry and a vocal advocate of sustainable mining practices on the international stage.

Dr. Nieto obtained his degree in mining engineering from the University of Guanajuato. During the early 1990s, he worked in underground gold and silver mines. Later, he received a scholarship from the Colorado School of Mines, where he earned his master's degree in 1995. His thesis research project focused on geostatistics, based on a lead characterization project in Leadville, Colorado. He was then invited by the Ecole des Mines de Paris in France to work on a research project in mining geostatistics, completing his second master's thesis dissertation in 1997. Upon returning to the United States, Dr. Nieto pursued a PhD in Mining Engineering, completing his thesis dissertation in 2001. His PhD research focused on developing a virtual reality real-time mining modeling and monitoring system, which was successfully tested at the Morenci Mine in Arizona.

After completing his academic training, Dr. Nieto served as a professor at Penn State University, where he taught several mining courses. He later accepted a position as Vice President of Global Engineering Education at Laureate Universities. Subsequently, he served as Chair and Professor at the University of the Witwatersrand in Johannesburg, South Africa.

Following his academic career, Dr. Nieto held various leadership positions in the mining industry, including Chief Technology Officer of The CRC Mining in Australia, Director of the Mining Tech Center, and Vice President of Technology and Innovation with Penoles Industries. Additionally, he served as Vice President of Innovation and Technology with Gold-Fields in Ghana, West Africa.

Currently, Dr. Nieto is the Director of the FLS Mining Technology and Research Center located in Salt Lake City, the United States. His current research interests include mining innovation and technology, mineral reserves, and mineral economics. Dr. Nieto also advises graduate students at several universities worldwide, including MBA teams at Harvard and Cambridge Universities, on topics related to 'the critical role that the mining industry will play in powering future economic growth' and 'The role of sustainable extraction and processing of critical minerals in driving the current tech energy transition'.

Dr. Nieto was inducted to the National Academy of Engineering, Mexico in 2018. He is a Qualified Professional (QP) accredited by The Mining & Minerals Society of America. You can find his latest publications and courses on his website at www.TonyNieto.com.

Introduction

The purpose of this book is to introduce a clear and comprehensive innovation model that can be applied to any type of resource-extractive industry, including mining. The innovation model proposed is designed to clearly indicate the operating trends, or innovation trends, that need to be considered today by the mining and minerals industry to achieve optimum operating conditions essential for any mining operation of the future.

Five key operational conditions have been carefully identified through years of research, teaching, and consulting for the mining industry. These five optimum mining operational conditions, or mining innovation drivers, are as follows: (1) Achieving maximum safety, (2) simplifying systems, (3) using smart-intelligent solutions, (4) designing stealth operations, and (5) following a sustainable strategy. The five innovation drivers are common denominators to any mining method, regardless of their operational nature or commodity being mined either today or in the future.

For instance, to achieve maximum safety, the mining industry can implement advanced safety technologies such as remote-controlled equipment and drones to eliminate risks associated with human presence in hazardous areas. To simplify systems, the industry can use automation technologies to reduce manual intervention in mining operations, resulting in increased efficiency and reduced errors.

Similarly, to use smart-intelligent solutions, the mining industry can incorporate artificial intelligence (AI) and machine learning technologies to enable predictive maintenance and real-time monitoring of mining equipment. Designing stealth operations can involve minimizing the impact of mining operations on the environment and communities by using sustainable mining practices. By following a sustainable strategy, the mining industry can start adopting renewable energy sources such as wind and solar power to reduce the carbon footprint of mining operations.

It is thus envisaged that the mining innovation model introduced in this book can serve as an initial guideline for the mining industry to better identify the current and future technologies that need to be addressed today. This will ultimately enable the industry to achieve the five operational conditions mentioned earlier to effectively operate the 'mine of the future'.

Note that illustrations provided in each of the technology sections within the chapters in this book are infographic ideograms – they are neither technical schematics nor engineering drawings. The illustrations provided in this book are included with the purpose of assisting the reader to easily visualize each of the technologies described and do not depict any specific tech brand or manufacturer.

Following is a quick description of the key chapters presented in this book.

Chapter 1: The dynamics of innovation, long-wave tech-business cycles

This chapter sets the foundation for the book by discussing the historical role of innovation and the economic development of any industry including mining. It also covers the challenges that the mining industry faces in the rapidly changing technological landscape. It establishes the need for innovation in the mining industry and introduces the main theme of the book.

Chapter 2: The information–innovation cycle

This chapter discusses the relationship between information and innovation in the mining industry. It introduces the concept of the information–innovation cycle and how it can be used to drive innovation in mining.

Chapter 3: Innovation driver of the industry of the future

This chapter covers the innovations that are driving the mining industry forward. It explores how these innovations are being used to improve mining efficiency, safety, and sustainability.

Chapter 4: Innovation constraints in mining

This chapter examines the barriers to innovation in the mining industry. It discusses the challenges that mining companies face in adopting new technologies and explores ways to overcome these challenges.

Chapter 5: The five innovation drivers for the minerals industry

This chapter identifies the five key operational conditions that are essential for driving innovation in the mining industry. It explores how these conditions can be created to support the mine of the future.

Chapter 6: The future of mining: safe

This chapter explores the future of mining safety. It examines the latest innovations in mining safety and how they can be used to create a safer mining industry.

Chapter 7: The future of mining: simple

This chapter looks at how mining can be made simpler through the use of technology. It explores the latest innovations in mining automation and how they are transforming the mining industry.

Chapter 8: The future of mining: smart

This chapter discusses the concept of smart mining and how it can be used to improve mining efficiency and sustainability. It explores the latest innovations in mining technology, such as the Internet of Things (IoT), and how they are being used to create smarter mines.

Chapter 9: The future of mining: stealth

This chapter explores the future of the mining footprint impact on the environment and society. It examines the latest innovations in mining technology that are making mining operations more discrete and less intrusive to the environment.

Chapter 10: The future of mining: sustainable

This chapter discusses the importance of sustainability in the mining industry. It explores the latest innovations in sustainable mining, such as renewable energy and water conservation, and how they can be used to create a more sustainable mining industry.

Chapter 11: Innovation roadmap to the mine of the future

This final chapter brings together the themes of the book and provides a roadmap for the future of innovation in the mining industry. It explores the key challenges and opportunities facing the industry and provides guidance on how mining companies can embrace innovation to create a more sustainable and successful future.

Chapter 1

The dynamics of innovation, long-wave tech-business cycles

Joseph Schumpeter, an economist professor from Harvard University, developed the theory of long-wave business cycles (Schumpeter, 1954), which suggests that world economies undergo cyclic periods of expansion and contraction driven by waves of technological innovation. Schumpeter's theory provides valuable insights into the relationship between innovation and economic development, highlighting the transformative power of technological advancements. Schumpeter's theory helps illustrate six historical innovation waves that have shaped the world economy from the Industrial Revolution to the present day. See Figure 1.1.

First Wave (1780): hydropower and textiles

The First Wave of innovation in the late 18th century was characterized by the adoption of hydropower and the rise of the textile industry. The invention of the multi-spindle spinning frame revolutionized the textile production, leading to increased efficiency and establishment of factories. These advancements marked the beginning of industrialization, paving the way for mass production and transformation of society.

Second Wave (1840): steam power and railroads

The Second Wave, which emerged in the mid-19th century, witnessed the widespread adoption of steam power and the development of railroad networks. Steam engines powered factories,

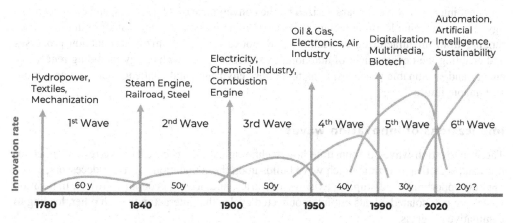

Figure 1.1 Innovation waves experienced by the global industry.
Source: Nieto (2019b)

DOI: 10.1201/9781032622699-1

enabling the production of goods on a larger scale and at a faster pace. The introduction of railroads revolutionized transportation, facilitating the movement of people and goods across vast distances, connecting previously isolated remote regions, promoting trade and economic growth.

Third Wave (1900): electricity and chemicals

The Third Wave emerged around the turn of the 20th century, with the advent of electricity and the rapid growth of the chemical industry. The widespread adoption of electricity transformed manufacturing processes, enabling factories to operate more efficiently, and enhancing productivity. The chemical industry, with innovations such as synthetic dyes and fertilizers, spurred advancements in agriculture and development of new materials, fostering further economic expansion.

Fourth Wave (1950): petrochemicals and electronics

The Fourth Wave emerged in the post-World War II period and was characterized by the rise of petrochemicals and electronics. The discovery and extraction of oil reserves led to the creation of a petrochemical industry, which revolutionized the production of plastics, synthetic fibers, and other materials. Simultaneously, the development of electronic technologies in the 1960s, such as transistors and integrated circuits, paved the way for the emergence of computers and the information age.

Fifth Wave (1990): digital transformation

The Fifth Wave, which emerged in the 1990s, marked the beginning of the digital era. The rapid advancement of computing power, the internet, and telecommunications technologies transformed various sectors of the economy. E-commerce, online services, and digital communication revolutionized the way businesses operate, fostering globalization and expanding opportunities for innovation and entrepreneurship on a global scale.

Sixth Wave (2020): AI, robotics, clean tech

The ongoing Sixth Wave is characterized by the convergence of AI, robotics, and clean technology. AI and machine learning technologies are driving automation and transforming industries ranging from manufacturing to healthcare. Robotics is revolutionizing production processes and enabling the development of autonomous systems. Clean technology including renewable energy and sustainable practices, aims to address environmental challenges and create a more sustainable future.

Implications of innovation waves

These innovation waves demonstrate how technological advancements have reshaped economies and societies over time. Each wave builds upon the foundations of its predecessors, creating new opportunities, disrupting existing industries, and driving economic growth. Innovation waves are not isolated events but interconnected cycles that interact with each other, leading to cumulative progress.

Innovation is a key driver of economic development, fueling entrepreneurship and creative disruption of old industries. Each wave of innovation brings about new opportunities for entrepreneurs to introduce disruptive technologies and business models, leading to the rise of new industries and the decline of obsolete ones. This innovation process generates economic growth, employment opportunities, and improvements in living standards.

The Sixth Wave, currently being experienced by society, highlights the increasing focus on sustainable technology and the urgent need to address environmental challenges. Clean tech innovations are essential for mitigating climate change, reducing carbon emissions, and transitioning to a more sustainable and resilient economy. AI and robotics play a crucial role in developing energy-efficient solutions, optimizing resource utilization, and enabling sustainable practices across various industries.

In 2020, the US National Academy of Engineering (NAE) published the results of a study aimed at identifying current and future global technology challenges. The NAE report identified several major technology challenges and classified them into 14 grand themes listed later.

Table 1.1 Key technologies intimately related to mining.

 1. Advance teaching & learning technology
 2. Secure cyberspace
 3. Improve solar energy
 4. Provide access to clean water
 5. Enhance virtual reality (VR)
 6. Provide fusion energy
 7. Reverse-engineering the brain
 8. Prevent terrorism (nuclear)
 9. Engineer medicine
10. Manage the nitrogen cycle
11. Advance health informatics
12. Develop carbon sequestration methods
13. Restore urban infrastructure
14. Engineer new tools for scientific discovery

These 14 challenges mentioned in the NAE report served as the guide for this book to define potential challenges the mining industry will face during the following years. The NAE report, for example, identifies several key technologies intimately related to mining such as energy, water, health, and safety.

However, even though the NAE report served well as an initial reference to help us define current and future technology trends, the report failed to recognize another important challenge our society will certainly face during the following decades: The sustainable supply of raw material, including the supply of minerals and metals; no new technology can be developed – not today, nor in the future – without the use of raw minerals.

Metals and minerals, only possible to be extracted and recovered by mining and metallurgical methods, are heavily used in every key technology such as electronics, computing, communication, transportation, batteries, robotics, aerospace, and so forth. In fact, metals and minerals are the key raw material used in every strategic industry and tech sector mentioned in the NAE report.

Table 1.2 Five grand tech challenges 2030.

1. Sustainable energy and water tech
2. Information tech
3. Health tech
4. Intelligent automation
5. Supply strategic raw minerals

Source: Nieto (2019a)

The Table 1.2 further groups the 14 NAE technology challenges into five grand tech challenges, including the supply of raw minerals, that the industry of today will most likely face during the following decades.

Industry in general is currently experiencing a tremendous digital and technology transformation, under the sixth innovation cycle currently being experienced by society. With the rapid advancement of technology any industry can fall victim to sudden disruptive technology shifts, potentially rendering any company obsolete in just a few years.

Schumpeter's theory of long-wave business cycles provides a valuable framework for understanding the relationship between innovation and economic development for the minerals industry. The six historical innovation waves, from hydropower and textiles to AI, robotics, and clean tech, exemplify the transformative power of technological advancements. These waves have reshaped economies, driven entrepreneurship, and laid the foundations for progress in various sectors including mining.

As we move forward, it is crucial to recognize the significance of sustainable technology and the need to harness innovation for the benefit of both society and the environment. Embracing the opportunities presented by innovation can foster economic growth, address pressing global challenges, and strive for a more sustainable and inclusive future.

To succeed during the 21st century, the mining industry needs to take innovation seriously and consider it an integral element of a modern corporate business plan.

Chapter 2

The information–innovation cycle

The conventional dictionary explanation of *innovation* is 'the use of better solutions that meet new requirements'. However, this definition is incomplete when considering innovation as an active element within the intricacies of the mining industry or any extractive industry per se. A revised definition of innovation, in the context of applied innovation within a complex industrial corporation, could be better defined as 'the outcome of ongoing activities that transform ideas to improve an existing condition, resulting in a higher incremental value'.

Innovation activities, or functions, within any given industry, can be many; however, within the context of designing a practical innovation model, the formulation can be based on these three phases: Creation, transfer, and enhancement, of information. In other words, as seen in Figure 2.1, innovation itself is the result of an information cycle that starts with data acquisition and data analytics which transform raw data into 'information'. By using science and research, information is then enhanced into forming conceptual models aimed to solve specific problems or to improve certain conditions, resulting in new solutions, for example, by improving a service, a system, or a product.

Once new solutions are adopted and fully implemented, a new generation of technology is established along with the creation and acquisition of new and enhanced raw data, thus commencing a new information–innovation cycle.

The infographic below describes the cyclic transformation process of data, into information, concept, and solutions, resulting in an innovation–technology cycle.

The cyclic information-innovation process involves the continuous transformation of data into new technology solutions, which generate more data, leading to the creation of new solutions, and so on. As indicated in the outer rim of the graph, the innovation cycle process consists of data transformation into information, then to the concept, and finally into a solution. Each of these transformational stages is based on four primary activities: Learning, research, testing, and innovation.

1. Analyze: Data is collected and analyzed in this first step to extract valuable information. Data analytics tools are used to transform raw data into meaningful insights, providing a better understanding of patterns and trends.
2. Enhance: Once information is extracted from the data, it is further enhanced through scientific research, resulting in the creation of new concepts and models. These models help in understanding complex systems and identifying potential solutions to problems.
3. Testing: The third step involves testing the concepts and models developed in the previous step. This helps evaluate their effectiveness and identify any flaws or limitations that need to be addressed.

DOI: 10.1201/9781032622699-2

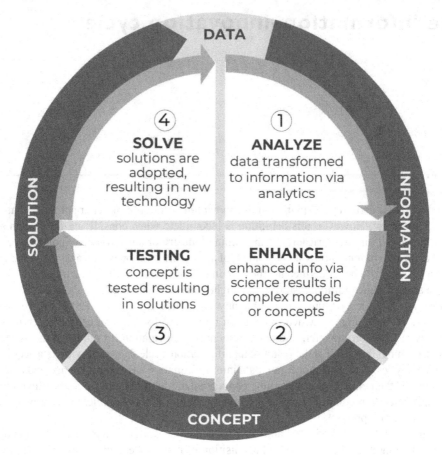

Figure 2.1 The information–innovation cycle.

Source: Nieto (2019b)

4. Solve: In the final step, new solutions are developed based on the tested concepts and models. These solutions are then adopted and integrated into the existing technology, resulting in the creation of new data, and completing the information–innovation cycle.

The information–innovation cycle is a continuous process that drives innovation and progress. By constantly analyzing and improving data, researchers and innovators can develop new technologies that meet evolving needs and solve increasingly complex problems.

Chapter 3

Innovation driver of the industry of the future

'Difficult to see, always in motion the future is' – a quote from *Star Wars* well describes the difficulty of foreseeing the future and how 'the mine of the future' will operate and what technologies will still be relevant and used at the operational level.

Nowadays, more than ever, production-driven industries such as mining are easily impacted by technology shifts that can result in a rapid loss of competitiveness. Thus, the continuous implementation of applied innovation is essential in any industry, including the mining industry, to stay relevant and competitive during the following decades. For the mining industry to succeed, innovating today is crucial for maintaining high levels of growth, productivity, and sustainability.

Mining technology has undergone a tremendous transformation over the past few decades, with advancements in digital technology, automation, and robotics revolutionizing the industry. Mining technology has evolved so much during the past 20 years that if we compare mining technology in the 1980s with today's technology, the differences are staggering. In the 1980s, mining operations were primarily based on analogical processes – all analytics were done by hand, operating heavy machinery without any sensorial systems. Without GPS or other modern navigation systems, miners had to rely on physical maps and compasses to navigate in open pit mining and underground tunnels and shafts. This made mining a time-consuming and labor-intensive process, with high risks of accidents due to the lack of safety features in the equipment.

Moreover, communication systems in the 1980s were limited, making it difficult for miners to communicate with each other and coordinate their work. The internet, Wi-Fi, and other digital technologies that we take for granted today did not exist back then.

Fast-forward to the present day, the mining industry has undergone a significant transformation, with the integration of modern technologies such as GPS, Wi-Fi, digital technology, VR, augmented reality (AR), and AI. These innovations have enabled miners to work more efficiently and safely.

GPS has made it easier to navigate through the mines, allowing miners to accurately locate their position and avoid potential hazards. In addition, the internet and Wi-Fi have made it possible to access real-time data and to communicate effectively among miners, and share data in real-time with equipment using mesh-wireless network systems. This has allowed for better collaboration and decision-making, leading to more efficient and productive mining operations.

Furthermore, the introduction of digital technology has led to the development of advanced monitoring systems that can track everything from equipment performance to environmental conditions. This has allowed miners to optimize their operations, reduce downtime, and increase safety.

DOI: 10.1201/9781032622699-3

The integration of VR and AR technology has also enabled miners to train and prepare for different scenarios and hazardous conditions. They can simulate emergency situations and learn how to respond to them, without exposing themselves to real-life dangers.

Finally, the introduction of AI and automation has revolutionized the mining industry, reducing the need for human labor, and moving miners from hazardous environments to remote operational centers increasing safety and efficiency. Advanced machines can now perform repetitive and dangerous tasks, allowing miners to focus on more complex and critical operations.

The mining industry has come a long way since the 1980s. The implementation of modern technologies such as GPS, internet, Wi-Fi, digital technology, VR, AR, and AI has transformed the mining industry, making it safer, more efficient, and productive. These advancements have made mining operations more sustainable, helping reduce the environmental impact of the industry while maximizing profits. However, it is imperative for the mining industry to have a clear idea of current and future operational trends to identify the technology that will be essential in the future and to prepare with a clear innovation strategy to be implemented at every stage of the value cycle within the company's business and growth plan.

As witnessed during the past decades, well-established companies such as Kodak from the photo industry, Blackberry from the communication industry, and Blockbuster from the media industry quickly lost their market relevance by failing to recognize the dynamics of new emerging technologies. The lack of an agile and clear innovation strategy and technology roadmap within their corporate business models most likely contributed to their sudden loss of technology relevance and eventual business collapse.

Developing an innovation strategy and technology roadmap for the mining industry is not an easy exercise. Mining operations vary significantly in terms of the mining methods used and the mineral commodity being extracted, facing very different operational challenges that require different technologies. Thus, the motivation to develop a mining tech innovation model, as introduced in this book, is to help forecast how the mine of the future will function during the following years.

Chapter 4

Innovation constraints in mining

The benefit of preparing a strategy for the mine of the future is obvious. Developing new technologies resulting from applied innovation is key to future success for any industry including the mining and minerals industry.

Innovation is crucial for the success of mining operations in all eight phases of the process, from extraction to tailings management; see Figure 4.1.

The mining industry is facing various challenges such as depleting reserves, increasing environmental regulations, and rising costs. Innovation can help address these challenges and improve efficiency, safety, and sustainability.

In the extraction phase, innovation can improve the accuracy and efficiency of mineral exploration, making it easier to find new mineral deposits. New technologies such as drone-based mapping and autonomous drilling can also make the extraction process safer and more efficient.

In the materials handling phase, innovation can improve the speed and efficiency of transporting materials from the mine to the processing plant. Autonomous trucks and conveyor belt systems can improve the speed and accuracy of materials handling while reducing labor costs.

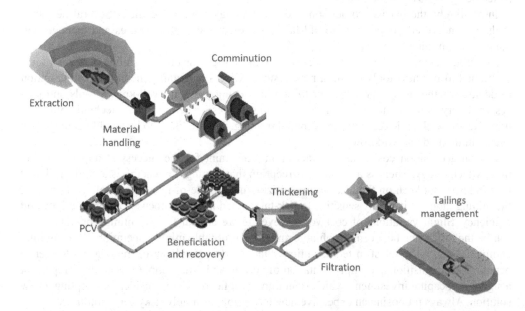

Figure 4.1 Typical mining and mineral process flowsheet. (FLS, MTRC Services, Public Domain)

DOI: 10.1201/9781032622699-4

In the comminution phase, innovation can improve the energy efficiency of crushing and grinding processes. High-pressure grinding rolls and vertical water-energy-efficient mills are examples of new technologies that can reduce energy consumption while improving mineral liberation.

In the pumps, cyclones, and valves (PCV) phase, innovation can improve the efficiency of pre-concentration processes, reducing the amount of waste material that needs to be processed. New sensor technologies can help identify valuable minerals in the ore stream, making it easier to separate them from waste.

In the beneficiation and recovery phase, innovation can improve the efficiency and sustainability of mineral processing. New processing technologies such as bioleaching and hydrometallurgy can reduce the environmental impact of mining while improving mineral recovery rates.

In the thickening phase, innovation can improve the efficiency of water management and reduce the risk of tailings dam failures. New technologies such as deep cone thickeners and high-rate thickeners can reduce the water content of tailings while improving their stability.

In the filtration phase, innovation can improve the efficiency and sustainability of dewatering processes. New technologies such as ceramic filters and vacuum belt filters can improve the quality of the filtrate while reducing water consumption and disposal costs.

In the tailings management phase, innovation can improve the safety and sustainability of tailings storage facilities. New technologies such as geosynthetic liners and remote monitoring systems can reduce the risk of environmental damage and improve the long-term stability of tailings dams.

By embracing new technologies and processes, mining companies can improve efficiency, safety, and sustainability while addressing the challenges facing the industry. Innovation is key to ensuring the long-term success of mining operations and the responsible management of mineral resources.

Thus, the question is why new technology is not being applied on a large scale in mining, and hence delaying its adoption rate.

Interestingly, the barriers to adopting new technology are not technical. Several new tech tools are underused, mainly due to old-fashioned organizational structures and management plans within mining companies.

Adopting new technologies can be expensive and time-consuming, as mine personnel must be trained to use new tools within a new system. Miners and staff generally have the option to decide whether to use new tech or to continue using previous well-known tools and processes. Every new technology requires a learning process. Therefore, new technology that is user-friendly and intuitive, requiring minimal training, is more likely to be quickly adopted and used within the daily workflow.

Capital acquisition cost is another important constraint for the successful implementation of new technology. There is a common perception that mining, in general, is an old-fashioned tech industry not keen on the idea of modernizing using state-of-the-art technology. In reality, the mining industry is highly sensitive to risk-taking and is greatly focused on productivity and efficiency. Slight variations of cost versus revenue are crucial for any mining operation and can be the difference between a profitable and an unprofitable mining operation. Thus, mining companies can only invest in tech solutions that clearly provide an operational cost-revenue benefit. The acquisition and implementation of current and future mining technology represent a significant capital investment, which is an important factor when considering adopting a new solution. Always proposing an expensive new technology is highly risky. Any mining company must be aware that the process of innovating mainly consists of trying and failing, several times, until a successful stage is achieved.

Clear communication between the mining client and tech suppliers is another potential constraint to a successful innovation process. Companies with tech offices purchasing new technologies often fail to properly communicate with operational onsite units. Conversely, mining units purchasing technology onsite often fail to communicate with the corporate tech office, resulting in short spans of effective use with no post-purchase training or maintenance plans. The same is true when implementing digital transformation projects in mining, a process often carried out without a communication plan at the mine site. Acquiring a tech tool is only the first step toward digital transformation. Both the tech provider and the mine client need to secure an onsite ongoing operational management plan with a dedicated team to ensure a successful adoption will be achieved.

Chapter 5

The five innovation drivers
for the minerals industry

The mining industry has been experiencing significant changes in recent years. With the ever-increasing demand for natural resources, mining companies are facing complex challenges in meeting production targets, while ensuring safety, reducing operational costs, and sustaining the environment. Innovation has become the key to unlocking the full potential of the mining industry.

Innovation in mining should not just focus on the introduction of new technologies, but rather on defining the ultimate operational conditions that the mine of the future will need to achieve. By understanding these conditions, mining companies can then identify the specific technologies required to achieve them.

As seen in Figure 5.1, the five ultimate operational conditions that the mine of the future should aim to achieve are as follows: Achieving maximum safety, simplifying systems, using smart-intelligent solutions, designing stealth operations, and sustaining an environmental-wellbeing economic model. The five innovation drivers have been defined as the result of several years of teaching, research, and consulting for the mining industry and are applicable to any type of mining operations: Metal or nonmetal, underground or surface, soft-rock or hard-rock mining. These five innovation drivers are thus common denominators to any mining industry regardless of their operational nature today and in the future.

Achieving maximum safety

Mining is a hazardous industry, and safety is of paramount importance. The mine of the future should be designed with safety as a top priority, using advanced technologies to mitigate risks and prevent accidents. By prioritizing safety, mining companies can not only protect their workers but also increase productivity, as a safer working environment leads to increased employee morale and job satisfaction.

Simplifying systems

Complexity in mining operations can lead to inefficiencies and increased costs. Simplifying systems can help reduce complexity, streamline processes, and increase productivity. By implementing simpler systems, mining companies can optimize their operations, reduce downtime, and improve the overall performance of their mines.

DOI: 10.1201/9781032622699-5

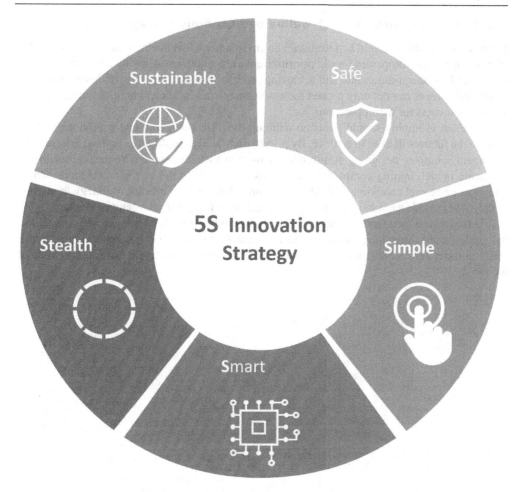

Figure 5.1 The 5S Innovation Strategy.

Source: Nieto (2019a)

Using smart-intelligent solutions

The use of smart-intelligent solutions can help mining companies optimize their operations by using data analytics and AI. These technologies can enable real-time decision-making, predictive maintenance, and improved operational efficiencies.

Designing stealth operations

Mining companies need to reduce their environmental footprint and minimize their impact on local communities. Designing stealth operations that are less disruptive to the environment and local communities can help reduce the negative impact of mining operations.

Sustaining an environmental-wellbeing economic model

The mine of the future should aim to sustain an environmental-wellbeing economic model. This means that mining companies should prioritize environmental sustainability and social responsibility in their operations, while still achieving profitability. By adopting sustainable practices, mining companies can not only protect the environment and local communities but also ensure long-term success for their operations.

Innovation in mining should focus on defining the ultimate operational conditions that the mine of the future will need to achieve. By prioritizing safety, simplifying systems, using smart-intelligent solutions, designing stealth operations, and sustaining an environmental-wellbeing economic model, mining companies can unlock the full potential of their operations. These ultimate operational conditions should be the foundation for any innovation strategy in mining, as they will provide a roadmap for the specific technologies required to achieve them.

It is thus envisioned that the innovation model introduced in this book can serve as an initial guideline for the mining industry to identify the technologies that need to be implemented today to ultimately achieve the five operational conditions to effectively operate the 'mine of the future'.

Chapter 6

The future of mining

Safe

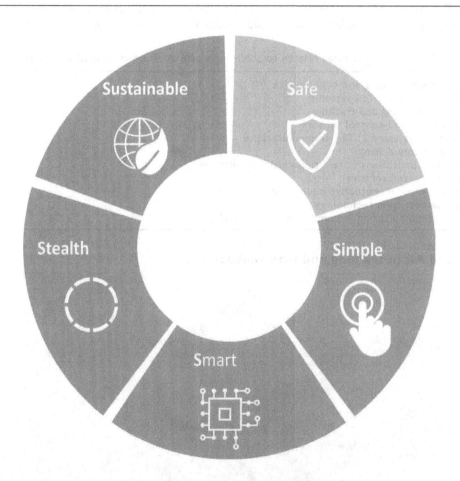

The first operational condition in mining to be achieved is *total safety*. The innovation model introduced here proposes the use of technology to achieve such a condition of total safety. This is also accomplished by creating innovation policies that work in tandem with the operation to achieve an ultimate safe operational condition through the whole value cycle within the mining operation. The mine of the future will focus on a very strict zero-accident/incident policy. Fatalities will be an extremely rare event due mainly to the heavy application of automated systems; VR and AR for safety training purposes will be extensively used, thus driving down incident and

DOI: 10.1201/9781032622699-6

accident rates. Besides automation, the use of remote-operated equipment, and advanced smart technology in mining equipment, is and will be a key tech trend that will significantly improve safety in mining.

The mining operation of the future will not just aim to reduce the number of accidents and incidents using state-of-the-art technology, but also invest in the use of noninvasive biosensor technology to continuously monitor the health and safety of every miner and staff to maintain and even improve the health of every employee. Using smart sensors and AI to continuously assess personnel health and stress levels – and, when necessary, promote healthier diets, exercise routines, meditation, and even suggest a well-deserved break from their mining shifts – the mine of the future will commit to improving the wellbeing of its employees.

Following is a short description of possible technologies that can be developed or implemented to promote and achieve a totally safe operational condition in mining in the near future.

Table 6.1 List of technologies related to achieving a safe operational condition in mining.

VR and AR in training and maintenance
Remote and automated systems
Real-time mapping and tracking
Real-time ground control sensors
Through-the-ground (TTG) communications
Rapid borehole drilling
Biosensor applications in real-time health monitoring
• Body-mounted sensors
• Smart personal protective equipment
• Fatigue detection technology
• Vibration monitoring in human-operated vehicles

VR and AR in training and maintenance

One of the defiant challenges in the mining industry is knowledge management. Expert operators acquire valuable knowledge about mining processes over the years; the transfer of this knowledge to new operators is a key issue. In this context, VR is used to provide experts and trainees with an immersive environment to better transfer knowledge and to quickly learn critical tasks under a steeper learning curve. VR is a technology for simulating mining process environments and situations with a high degree of realism and interactivity. Its training tools guarantee a simulated exposure to real-world working conditions, without the associated risks; the goal being to transfer the knowledge acquired in a virtual environment to a real one.

Currently, VR is being successfully applied in the mining industry to improve accident prevention and interactive training, through data visualization, machinery and equipment maintenance, accident reconstruction, simulation applications, risk analysis, and hazard awareness.

Augmented reality (AR) is one of the critical technologies for the transformation of the mining industry, in terms of optimizing the cost of maintenance and training. AR is indeed changing the future of mine safety through immersive learning. Both training and maintenance, and repair of equipment, machinery, and entire systems (e.g., conveyors, complete lifting systems, pipelines) can be carried out more efficiently by using AR technologies, to create engaging learning experiences, that drive meaningful learning outcomes and improve knowledge holdings.

Remote and automated systems

Remote-operated equipment and vehicles have brought significant advancements to the mining industry, particularly in terms of safety and productivity. This technology will continue to improve during the following years allowing for the operation of mining equipment and vehicles remotely and from a safe distance, reducing the risks associated with mining operations and increasing productivity.

One of the most significant benefits of remote-operated equipment is improved safety. By using remote-operated equipment, miners are no longer required to work in dangerous areas, such as underground mines or near dangerous machinery. Remote-operated equipment also eliminates the need for miners to physically touch or manipulate hazardous materials, such as explosives or toxic chemicals, further reducing the risk of injury or illness.

Another safety benefit of remote-operated equipment is the ability to quickly respond to emergencies. With remote control, miners can quickly shut down equipment or vehicles in case of an emergency, preventing potential injuries or damage to the equipment. Remote operation allows for better monitoring of equipment and vehicles, identifying potential safety hazards before they become serious issues and improving productivity.

In terms of productivity, remote-operated equipment and vehicles have numerous benefits. Remote operation allows for more efficient use of time, as miners can operate multiple pieces of equipment simultaneously, reducing the need for additional labor.

Remote-operated equipment also allows for mining operations to continue during times of inclement weather or hazardous conditions.

While the initial investment may be higher, remote-operated equipment improves safety and labor costs. With remote-operated equipment, mining companies can improve their operations while keeping their employees safe and productive.

Real-time mapping and tracking

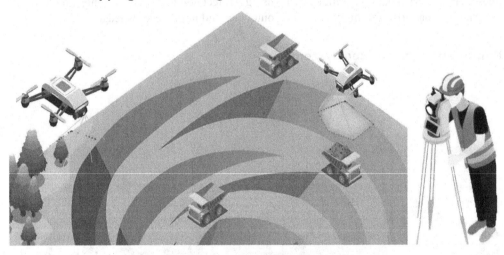

Successful mining operations depend on intelligent and measurable processes, which can be customized using data and analysis to optimize and monitor process execution, depending on the latest situational context and information. This high degree of automation, along with the incorporation of AI, is driving the implementation of analytical technologies for real-time tracking and monitoring. Some real-time tracking and monitoring systems use active Radio Frequency Identification (RFID), sensors, wireless mesh network, active GPS-based collision avoidance warning systems, and radio frequency proximity detection.

Other types of systems are wireless and automated to provide real-time tracking of miners and mining equipment and monitoring of environmental conditions (i.e., emission of toxic gases such as methane and carbon monoxide) within the mine. In addition, they provide audible warnings to prevent collisions between mining vehicles and send warnings to alert the miner who is approaching unsafe areas. These systems also allow the miner to send an emergency message to the surface control station. Usually, these systems allow for monitoring of personnel access levels, monitoring of personnel positioning in the mine, working time control, emergency personnel notification, and emergency personnel detection and evacuation.

High accuracy 3D mapping is crucial in working conditions. Today, the wireless positioning system can track people and objects with an accuracy of about half a meter. Such systems use small mobile tags attached to vehicles, or mine workers, along with a series of reference nodes placed in known locations around the mine being monitored in real time. Mine surveying and mapping using drones and laser technology makes it possible to constantly maintain control of the state of the excavations inside the mine, especially tunnels, pillars, chambers, and stopes. The 3D mobile laser scanners can be attached to mining vehicles or drones to produce real-time 3D point cloud data along the way, as the vehicles move through the mine.

Real-time ground control sensors

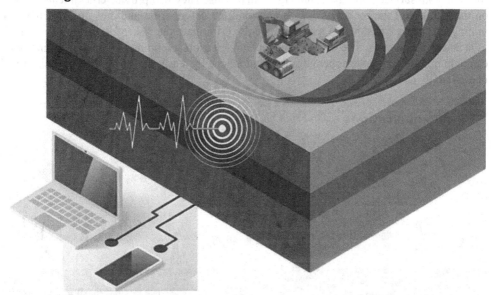

Ground control sensors are an essential tool in both surface mining and underground mining. They are used to monitor the stability of the rock mass and ensure the safety of miners and equipment. Thus, the use of ground control sensors and real-time communication technology in both surface and underground mining are of vital importance in ensuring the success of mining operations.

In surface mining, ground control sensors are used to monitor the stability of slopes, walls, and other structures. These sensors measure the deformation and movement of the rock mass and alert the mining engineers about any potential instability. This information is used to develop plans to mitigate the risk of rockfalls, landslides, and other hazards. Ground control sensors are also used to monitor the subsidence of the ground due to mining activities, which is crucial for maintaining the integrity of structures on the surface.

In underground mining, ground control sensors play a crucial role in ensuring the safety of miners and equipment. These sensors are used to monitor the stability of the surrounding rock mass and identify any potential hazards, such as roof falls or collapses. By monitoring the deformation and movement of the rock, mining engineers can determine where to place supports to ensure the safety of miners and equipment. Ground control sensors are also used to monitor the subsidence of the ground above the mine, which can impact the stability of the underground workings.

There are various types of ground control sensors used in mining, including extensometers, inclinometers, and borehole pressure cells. Extensometers measure the deformation of rock, while inclinometers measure the angle of the rock mass. Borehole pressure cells measure the pressure of water or gas in the rock, which can indicate the potential for rock failure. These sensors are typically installed in boreholes, which are drilled into the rock mass at various depths and angles.

In addition to ground control sensors, mining companies also use other technologies, such as radio detecting and ranging (RADAR), light detection and ranging (LiDAR), and drones, to monitor the stability of the rock mass. These technologies provide real-time data on the movement of the rock and can be used to develop more accurate models of the rock mass.

Thus, the development of real-time ground control sensors is an essential technology in the future in both surface mining and underground mining. They will provide crucial real-time data on the stability of the rock mass. By monitoring the deformation and movement of the rock, mining engineers can develop safety protocols to mitigate the risk of hazards such as rockfalls and roof collapses. The use of ground control sensors, combined with other technologies, can help mining companies operate safely and efficiently, ensuring the success of their mining operation.

TTG communications

TTG communications technology has played a crucial role in revolutionizing underground mining operations over the years. This technology allows for reliable communication between underground miners and surface personnel, enabling real-time monitoring of mining activities, and improving safety and efficiency.

The development of TTG communications technology in underground mining dates back to the mid-20th century when mining companies started experimenting with various methods of transmitting signals through the earth. The earliest methods were based on radio waves, which had limited success due to the high attenuation of radio signals through the ground. In the 1960s, the use of inductive coupling technology began to gain popularity. This technology involved the use of a wire loop around a borehole to generate a magnetic field that could transmit signals through the ground. However, this technology was still limited by the relatively short range of the signal and the need for careful borehole placement.

In the 1980s, the development of magnetic-field-based TTG systems began to show promise. These systems used magnetic fields to transmit signals through the ground and were more reliable than previous systems. However, they still faced issues related to signal interference and the need for complex infrastructure.

Over time, advancements in digital signal processing and electromagnetic modeling allowed for the development of more advanced TTG communication systems. These systems use low-frequency electromagnetic signals that can penetrate the earth's crust and travel long distances. They are also resistant to interference from other signals, making them more reliable than previous technologies.

The development of TTG communication technology has had a significant impact on underground mining operations. With real-time communication between underground miners and surface personnel, it is now possible to monitor mining activities and respond quickly to any potential hazards or emergencies. This has greatly improved the safety of miners and increased the efficiency of mining operations. Additionally, TTG communication technology has assisted with the advancement of automation and remote control in underground mining.

Thus, the development of TTG communication technology will continue to play a crucial role in the evolution of underground mining operations. Through advancements in electromagnetic modeling and digital signal processing, it will be possible to reliably communicate through the ground over long distances, enabling real-time communication between underground miners and surface personnel. This technology has the potential to improve safety in mining operations and has also opened new possibilities for automation and remote control. As technologies continue to evolve, it is likely that TTG communication systems will continue to play a vital role in the mining industry.

Rapid borehole drilling

Access to underground areas is an essential requirement for successful mining operations in the future. Historically, the process of drilling boreholes for access to underground sites has been slow and cumbersome, posing safety and productivity challenges in any mining operation, particularly in underground mining. The development of rapid borehole drilling technology will improve safety in the mining industry by enabling faster access to underground areas.

Rapid borehole drilling technology in underground mining has been evolving for several decades. Initially, drilling was performed using pneumatic or hydraulic drills that were slow and limited by the depth of the hole they could drill. Over time, the development of more advanced drilling methods and tools has improved the efficiency and safety of the drilling process.

One of the most significant advancements in rapid borehole drilling technology has been the introduction of directional drilling. Directional drilling allows for the drilling of boreholes at angles and in different directions, making it possible to reach underground areas that would otherwise be inaccessible. Directional drilling is particularly useful in cases where it is necessary to avoid sensitive areas, such as water or gas reservoirs.

In addition to the drilling process itself, improvements in communication technology have made it easier and safer to communicate between the surface and underground areas during drilling. Wireless communication technology has enabled miners to communicate more effectively, providing real-time updates on the progress of drilling and alerting the surface team to any potential issues.

The development of rapid borehole drilling technology will have a significant impact on the mining industry. With faster and safer access to underground areas, miners can work more efficiently and safely. Rapid borehole drilling technology will also enable more accurate exploration of underground areas, resulting in increased mineral reserves and improved mine planning.

Thus, the development of rapid borehole drilling technology represents an important challenge in the future of the mining industry. With the introduction of more reliable, sensorial, and directional drilling, access to underground areas will become faster and more efficient. The ability to communicate wirelessly during and after the drilling process will also improve the safety of miners and enable to regain real-time communications in case of an emergency. As technology continues to evolve, borehole drilling technology will play a critical role in safety in the mining industry.

Biosensor applications in real-time health monitoring

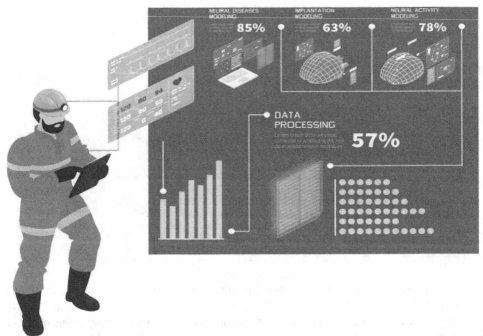

There is a great potential for the use of biosensors in mining operations, their application ranging from monitoring a variety of physiological data to early diagnosis and disease prevention, to facilitating miners' health management.

Biosensors allow mine workers to monitor their health status in real time; they are used by physicians for early diagnosis and prevention of worker illness. A biosensor is a portable, real-time, wireless health monitoring system. Current applications of real-time health monitoring biosensors include body temperature, heart rate, electrocardiogram, electroencephalography, electromyography, blood pressure, and glucose level, among many other bioindicators.

Although there are not yet commercial embedded-sensor-clothing applications in the mining industry, in particular, the usual safety vests used by miners can be transformed into a kind of smart clothing when biosensors are used. Thus, it is now feasible to embed miners' safety vests with both situational-awareness sensors (e.g., carbon monoxide sensors as well as sound, humidity, dust, and temperature sensors to monitor dangerous conditions in real-time in the area) and biosensors (e.g., body temperature sensor and pulse rate sensor to monitor the miner's primary physiological conditions), which could allow in the near future real-time measurement of both working conditions and mine workers' health.

Body-mounted sensors

With the advent of miniaturized sensor technology that can be used on the body, it is now possible to collect and store data from mine workers' movement working in an underground or a surface mine. This technology has the potential to be used in automated activity-profiling systems, producing a continuous record of activity patterns over extended periods of time. Such automated activity profiling systems rely on classification algorithms that can effectively interpret body sensor data and identify different activities.

Body-mounted sensor devices are equipped with the means of measuring a worker's pulse rate, amount of exercise, and emergency notifications. These devices monitor worker safety in mines for risk factors such as accidents from accidental contact, falling equipment, and inhalation of toxic gas. These devices are also capable of accurately determining a worker's safety.

A summary list of the technologies related to achieving a total safety condition in mining is as follows.

Smart personal protective equipment

For the mining industry, human safety is the top priority; it is the most important key performance indicator in any modern mining operation. Today, smart technology such as that found in smartphones and smartwatches has been integrated into personal protective equipment (PPE). PPE can connect to the mine communication infrastructure, such as underground WI-FI or long-term evolution (LTE) network, to provide real-time information on worker safety. PPE not only fulfills its original function of protecting workers from hazards but also collects data on worker functions, adjusts to internal and external conditions, and notifies workers of dangerous maneuvers and/or positioning. In addition, PPE can monitor and send warnings to the miners via their rugged smartphones or tablets, by providing real-time data, location-specific warnings, health monitoring, and improved communication. Future miners' PPE may include smart safety glasses with video monitoring, and capable of using AR, smart hard hats, biosensor-vests and clothing, smart rugged phones, and smartwatches that can monitor and inform mine personnel of potential safety risks in real time.

Fatigue detection technology

Monitoring the awareness levels of drivers and operators of heavy machinery in mines is vital. Fatigued operators are a danger to their safety and that of those around them.

They also reduce productivity and can cause damage and unnecessary wear to vehicles and other equipment. New technologies used for fatigue detection are based on electrophysiological monitoring of electrical activity in the brain; thus, it is not necessary to monitor the operator's eyes to establish his waking state. Several Canadian, Australian, and South African mining companies have equipped their employees with the technologies, such as Smart Caps, which are designed to monitor brain waves in order to measure fatigue.

Another technological alternative is based on platforms that use AI to detect driver fatigue. The platform connects to video cameras mounted on the truck's cabin constantly monitoring the driver's eyes for fatigue behavior. The system can also use vehicle's GPS, acceleration, and speed data, correlating fatigue data from other equipped sensors such as smart belts, and correlates it with factors such as the duration of a given work shift.

The system suggests an action plan to mitigate risk with short- and medium-term tasks and objectives, and then monitors the implementation of the recommended plan. The platform can also use other data sources such as vital sign trackers, sleep sensors, and sleep bracelets.

Vibration monitoring in human-operated vehicles

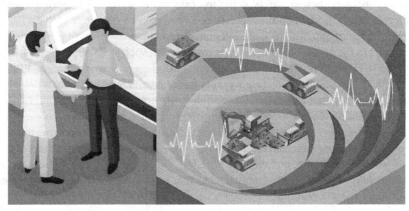

Heavy machinery produces whole-body vibration (WBV) and mechanical shock exposure to equipment operators when operating on rough surfaces and in harsh conditions. Mobile equipment operators in surface and underground mines are exposed to WBV. Innovative WBV measurement techniques are now being developed using WBV monitoring software to gather comprehensive data on WBV, in mining equipment during normal operation, to identify the main sources of exposure to dangerous WBV, and to explore possible control measures.

The development of applications for WBV monitoring will offer the opportunity to routinely monitor long-term data on WBV exposure. This information will be correlated with the activity being performed, the operating conditions, and the characteristics of the equipment.

The monitored information will allow the identification of appropriate control measures, and evaluation of the effectiveness of those measures, as part of a vibration risk management program.

Chapter 7

The future of mining

Simple

Simplicity is the second operational condition to achieve in the mine of the future. The mine innovation model proposes a simplification strategy when considering the development and adoption of new technology by promoting 'innovation by simplification' versus 'innovation by complexity'. Thus, simplification is a key tech driver when designing the mine of the future – looking for simplification at every stage and level within the value cycle of the mining system.

One key technology in the simplification process is the adoption of a modular system strategy. Modular operations in mining, in terms of production capacity, along the whole value-production cycle, are an important technology-driven strategy that can prepare any mining operation to quickly adapt to sudden market shifts that very often push a mining operation to either reduce or increase production capacity without incurring significant capital and operational costs.

Hence, modular systems are an excellent operational approach to stay simple and flexible, adapting to sudden production shifts as the mine matures from early stages of development to the extraction phase, and finally to the closure and reclamation phase.

DOI: 10.1201/9781032622699-7

Technology innovation indeed is a great strategy that can bring solutions to challenging problems; however, if not careful, new solutions can result in more complex systems that eventually will bring more problems. Innovation by complexity inevitably brings more systems and subsystems, thus more uncertainty, and more risk, which will eventually result in higher costs. Innovation by simplification is thus key to avoiding unintended complexity when solving problems or improving products or systems in mining. Some of the most common operational management systems used by the mining industry today deal with the identification of production constraints, operational stage identification, variation minimization, and simplification (Stamm et al., 2019).

Following is a description of current and future methods and technologies related to creating a simplified operational condition in the mine of the future.

Table 7.1 List of methods and technologies to achieve a simplified operational condition in mining.

Modular mining systems
Quality control and quality management
Productive maintenance
Theory of constraints (TOC)
Lean production
Six Sigma
Agile
Stage gate
Multi-attribute utility theory (MAUT)

Modular mining systems

The mining industry is known for its high capital intensity, long lead times, and high risks associated with project development. Therefore, it is crucial to employ an effective production capacity strategy that can mitigate risks and minimize capital costs. A modular, phased production capacity strategy can be an excellent approach to achieving these goals.

Modular production capacity involves the construction of smaller, self-contained production or processing modules that can be easily expanded or reduced based on the changing project needs. This approach allows for flexible and agile production planning, which can adapt to market conditions and mitigate risks associated with the volatile commodity prices commonly experienced in mining. The phased approach involves breaking down a large-scale mining project into smaller, more manageable production phases that can be completed independently. This approach ensures that each phase of the project is fully operational before progressing to larger production capacities, minimizing the risks of costly delays and potential project failures.

The production capacity modular approach enables mining companies to achieve cost savings by avoiding the construction of large, capital-intensive processing plants that may not be fully utilized during the first years of development. By building smaller, more flexible production modules, mining companies can match production capacity with market demand, reducing overcapacity and improving overall production efficiency. In addition, modular construction allows for the use of prefabricated components, reducing the time and cost of on-site construction, and minimizing environmental impacts.

The phased approach also helps mitigate risks and reduce capital costs. By completing each phase of the project independently, mining companies can ensure that each phase is fully operational and financially viable before proceeding to the next phase. This approach reduces the risks of costly project delays and potential project failures. Furthermore, by completing each phase independently, mining companies can reduce their initial capital requirements, spreading out the project's financial risk over time.

Another advantage of modular, phased production capacity strategies is their ability to improve project financing options. By completing smaller, self-contained production modules, mining companies can seek financing for each phase of the project separately, reducing their overall financial risk. This approach makes the project more attractive to lenders, as it reduces the amount of capital required upfront and increases the predictability of cash flows.

A modular, phased production capacity strategy, thus, is an effective approach for mitigating risks and reducing capital costs in mining projects. This approach allows mining companies to build smaller, more flexible production modules that can be easily expanded or reduced based on market conditions. The phased approach ensures that each phase of the project is fully operational before proceeding to the next phase, reducing the risks of costly project delays and potential project failures.

Quality control and quality management

Total quality management is defined as a customer-oriented process that aims at continuous improvement of processes in the mining industry. Total quality management in mining ensures that the work of miners and staff is directed toward the continuous improvement of the quality of a product or service, as well as improving the productivity of mining processes.

The emphasis on quality control should be placed on fact-based decision making with the use of key performance indicators and metrics to monitor progress. Total quality control is an integrated effort across the whole value cycle toward performance improvement at all levels within the mining system oriented at satisfying operational objectives such as quality, cost, scheduling, and workforce. It is expected that these activities will lead to achieving greater operational quality resulting in better customer satisfaction.

Total quality control, and total quality management, focus on customer satisfaction, a concept that is gaining significant traction as being championed by Amazon (the company) and other tech industries heavily focused on defining business and innovation strategies based on customer satisfaction.

Productive maintenance

| AUTONOMOUS MAINTENANCE | PREVENTATIVE MAINTENANCE & EARLY MANAGEMENT | QUALITY MANAGEMENT & ADMINISTRATIVE WORK | EMPLOYEE TRAINING & SAFETY |

Total productive maintenance is a strategy that eliminates the distinction between operation and maintenance roles by placing a strong emphasis on training working operators, or miners, to assist in the maintenance of mine machinery and equipment. The responsibility of carrying out maintenance does not rest solely with one department, or group of people within the mine, but depends on every worker who is operating and working with the mine equipment.

Productive maintenance aims to improve productivity by reducing unplanned downtime and asset availability, ensuring that mine and plant equipment are always available. This approach utilizes the skills of all workers within the mine and aims to incorporate maintenance into daily performance while emphasizing proactive and preventive maintenance practices.

When maintenance is a part of everyone's daily responsibilities of their job, and actively contributes to productivity, maintenance costs are significantly reduced, availability and profitability of equipment are increased, and also, as a result, there is improved teamwork and employee involvement.

Theory of constraints

Theory of Constraints (TOC) is a management methodology that emphasizes identification and management of bottlenecks, or constraints, in a system in order to improve overall efficiency

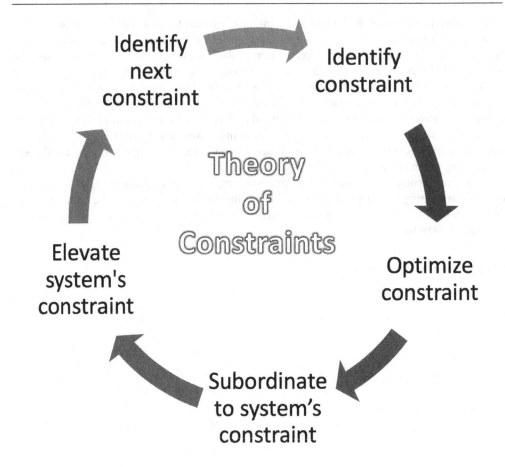

and productivity. In recent years, TOC has been adopted by several production-based industries including mining to drive improvements in both strategic and tactical planning.

In the mining industry, where operations are often complex and involve numerous processes, applying TOC can yield significant benefits since it provides a structured approach to identifying and addressing production and operational constraints. By using the TOC methodology, a mining company can systematically identify bottlenecks in the production process and focus resources on removing them. This can help improve the overall efficiency of the operation, reduce downtime, and increase production capacity.

TOC is based on the premise that in any operation, including a mining operation, there is at least one constraint or bottleneck that controls the rate at which profits are generated. TOC states that if this constraint or restriction is addressed first, the greatest improvement, in terms of profitability, will be achieved. In contrast, improvement projects based on unrestricted areas, or bottlenecks, are unlikely to provide significant improvements.

Removing this given restriction will, in practice, improve the overall mining operation process, and thus help mining companies gain a competitive advantage by responding in time to customer demands. In addition, reducing unnecessary work in the life cycle of the operation will result in lower costs and increased profitability. The central benefit of TOC is the reduction of inventory and operating expenses while increasing productivity.

One of the key advantages of TOC in mining is its ability to improve production scheduling. By identifying constraints and bottlenecks, TOC can help companies optimize their production schedules and ensure that resources are allocated where they are most needed. This can lead to a more efficient use of equipment and personnel, reducing the likelihood of idle time and increasing overall productivity.

Another advantage of TOC is that it promotes a focus on the overall system rather than individual components. In mining operations, it can be easy to become fixated on individual processes or equipment and lose sight of the bigger picture. By emphasizing the need to identify and manage constraints in the entire system, TOC encourages a holistic approach to production that takes into account the interdependence of different processes.

In addition, TOC can help mining companies better manage their inventory levels. By identifying constraints in the production process, companies can better understand where bottlenecks are likely to occur and adjust their inventory levels accordingly. This can help reduce the amount of excess inventory that needs to be stored and can also help ensure that the right materials are available when needed.

TOC can also help mining companies improve their quality control processes. By identifying constraints in the production process that are likely to impact quality, companies can take steps to address them before they become a problem. This can help improve the overall quality of the product and reduce the likelihood of customer complaints or rework.

At last, TOC can help mining companies improve their financial performance. By improving production efficiency, reducing downtime, and improving quality control, companies can reduce their operating costs and increase their revenue. This can lead to improved profitability and a stronger financial position.

TOC, if properly applied, can provide a powerful framework for improving the efficiency and productivity of mining operations. By identifying and managing bottlenecks in the production process, companies can optimize their production schedules, reduce downtime, improve quality control, and enhance their financial performance. For mining companies seeking to improve their operations and stay competitive in this rapidly changing industry, TOC is a valuable tool that should not be overlooked.

Lean production

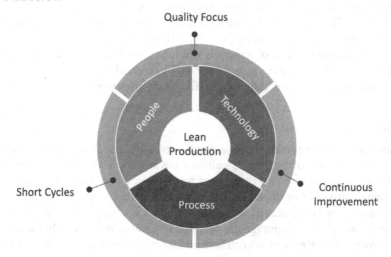

Lean production, or manufacturing, is a methodology that seeks continuous improvement, and the elimination of activities that do not add value to the production system; it involves all the mine staff to achieve this. Similarly, administrative tasks must be organized and optimized intelligently. Waste is understood as overproduction, waiting periods, unnecessary movement, unnecessary processes, inventory, defects, rework, as well as underutilization of people's capacity.

Lean production consists of a series of principles, concepts, and techniques designed to eliminate waste and establish an efficient and just-in-time production system, which allows internal or external customers, when applicable, to be delivered the expected quality of products when required, in the required quantity, and in the required sequence.

The lean production concept has proved to be successful in many industries, including mining. Implementing a lean production system in a mining operation can bring several benefits to the people, processes, and technology involved in the operation.

People

One of the most significant benefits of implementing lean production in a mining operation is the positive impact it has on the people involved. Lean production emphasizes the importance of employee involvement and empowerment. Employees are encouraged to identify inefficiencies in the production process and suggest improvements, which can increase their job satisfaction and engagement. This can also lead to a more collaborative and innovative work environment, as employees are given the freedom to experiment with new ideas and test new methods.

Furthermore, lean production can improve safety in a mining operation. By eliminating waste and standardizing processes, lean production can help reduce the risk of accidents caused by human error. This is particularly important in mining operations, where safety is a top priority.

Process

Lean production can bring several benefits to the processes involved in a mining operation. One of the most significant advantages is the reduction of waste. Waste can take many forms in mining, including excess inventory, overproduction, and defects. By identifying and eliminating waste, lean production can streamline processes and reduce costs.

Another benefit of lean production is the ability to improve lead times. In mining, lead times are critical, as delays can result in lost production and revenue. Lean production can help reduce lead times by improving the flow of materials and information throughout the operation. This can be achieved by implementing just-in-time (JIT) production, which involves producing goods only when they are needed, rather than building up inventory.

Technology

Finally, implementing lean production in a mining operation can bring several benefits to the technology involved. One of the most significant advantages is the ability to automate processes. Automation can help reduce costs, improve efficiency, and reduce the risk of accidents caused by human error. For example, automated machinery can help improve the accuracy of drilling and blasting, reducing the risk of accidents and improving production efficiency.

Lean production can also encourage the adoption of new technologies. By focusing on continuous improvement, lean production can help mining operations stay up to date with the latest technologies and equipment. This can help improve the operation's efficiency and competitiveness in the industry.

Implementing lean production in a mining operation can bring several benefits to the people, processes, and technology involved. By empowering employees, reducing waste, improving lead times, and adopting new technologies, mining operations can become more efficient, safer, and competitive in the industry. Lean production can be successfully applied in the mining industry to eliminate waste, and increase operational reliability, through the commitment of all personnel.

Six Sigma

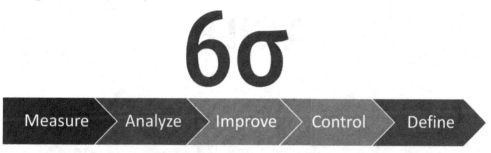

Six Sigma is a process improvement methodology that has been widely adopted in various industries, including mining, to simplify and improve operations. The mining industry has unique challenges that can be addressed through the application of Six Sigma principles.

Six Sigma is a data-driven approach that uses statistical analysis and other analysis tools to identify and reduce uncertainty and variability in a process. In mining, this approach can be applied to various processes, such as drilling, blasting, hauling, and processing, to improve efficiency, reduce costs, and increase productivity. Six Sigma helps identify waste, reduce variability, and improve the quality of output. The goal is to achieve a process that is capable of producing high-quality results consistently.

One of the key benefits of Six Sigma in mining is its ability to identify the root causes of problems. Mining operations are complex, and it can be difficult to identify the source of an issue. Six Sigma tools, such as the fishbone diagram, can help pinpoint the root cause of a problem by breaking down the factors that contribute to it. Once the root cause is identified, a targeted solution can be implemented to solve the problem, reducing the likelihood of it occurring again.

Six Sigma can also be used to optimize equipment performance. Mining equipment is expensive, and downtime can be costly. By implementing Six Sigma principles, equipment performance can be monitored, and issues can be identified before they become major problems. Predictive maintenance can be scheduled, reducing downtime and increasing equipment availability.

Six Sigma also helps improve safety in mining operations. Safety is a top priority in mining, and Six Sigma principles can be applied to identify potential hazards and develop solutions to mitigate them. By analyzing data and identifying trends, risks can be minimized and accidents can be reduced.

Mining companies that adopt Six Sigma principles can achieve a competitive advantage, creating a sustainable future for the industry.

Agile

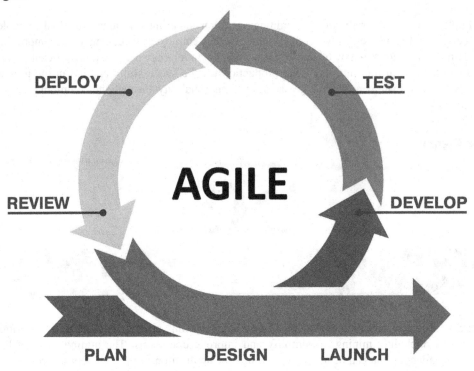

The Agile management strategy is a project management methodology that emphasizes flexibility, collaboration, and continuous improvement. It can help mining projects simplify operations in several ways.

1. Agile strategy is based on an iterative management process that is carried out in highly autonomous and self-organized teams capable of adapting to changing circumstances.
2. Agile methodologies prioritize delivering value to customers and stakeholders, rather than focusing on completing tasks or meeting deadlines. In the context of mining projects, this can mean focusing on extracting the most valuable ore first, rather than simply following a predetermined mining plan. This approach can help simplify operations by reducing the complexity of the project and focusing on what is most important.
3. Agile strategy creates organizational capacity to renew itself, adapt, and quickly recognize the opportunities that arise during the development and/or operational stages within a mining operation with the main objective of constantly adapting to external unpredictable factors, thus improving its effectiveness and productivity within different conditions and situations.
4. Agile methodologies also involve breaking a project down into smaller, more manageable pieces, and working on these pieces in iterations or sprints. In the context of mining projects, this can mean focusing on a specific section of the mine at a time and iterating on the mining process to optimize efficiency and productivity. By working in smaller iterations, mining projects can simplify operations and make it easier to identify and fix issues as they arise.
5. Agile methodologies emphasize on cross-functional teams, and collaboration and communication between different parts of the project team. In the context of mining projects, this can

mean bringing together experts from different areas such as geology, engineering, and operations to work on specific aspects of the project. By working collaboratively, mining projects can simplify operations and ensure that everyone is working toward a common goal.

6. Agile methodologies emphasize the importance of continuous improvement and learning from experience. In the context of mining projects, this can mean using data and feedback to refine the mining process and optimize operations. By continually improving and refining the mining process, mining projects can simplify operations and increase efficiency over time.

Agile, thus, as a management strategy can help mining projects to simplify operations by focusing on delivering value, working in smaller iterations, promoting cross-functional teams, and emphasizing continuous improvement. By adopting an Agile strategy approach, mining projects can simplify operations, increase efficiency, and improve overall project outcomes.

Stage gate

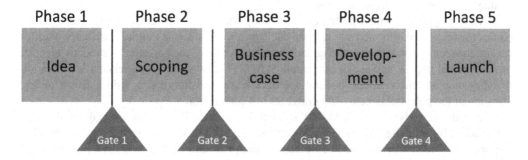

The stage-gate management strategy is a project management approach that involves breaking down a project into stages or phases and assessing each stage before progressing to the next one. This approach can be beneficial in mining projects as it allows for a more structured and systematic way of managing complex operations.

Some key aspects of a stage-gate management strategy that can help mining projects simplify operations consist of breaking down the project into distinct stages or gates, each with clear criteria for success or failure, so that the project team can focus on meeting specific goals and milestones. This helps simplify operations by providing a clear roadmap for the project's progression.

Each gate in the stage-gate approach includes a thorough evaluation of the project's progress to identify the risks and issues that may arise later in the project. Early identification of potential issues can help simplify operations by allowing the project team to address problems before they become more significant and costly.

Furthermore, the stage-gate approach involves a rigorous evaluation process for each gate, including a decision to proceed or stop the project. This helps simplify operations by ensuring that decisions are made based on objective criteria, rather than subjective opinions.

The stage-gate approach also involves regular reviews and updates of the project's progress, providing opportunities for improved communication between the project team and stakeholders. This can help simplify operations by ensuring that everyone is aware of the project's status and any issues that may arise.

Overall, a stage-gate management strategy can help mining projects simplify operations by providing a clear roadmap for the project's progression, identifying potential risks early, streamlining decision-making, and improving communication between the project team and the stakeholders.

Multi-attribute utility theory (MAUT)

Mining is a complex industry that needs to constantly innovate to bring new solutions to enhance operational efficiency, sustainability, and safety. In the quest for innovation, selecting the most suitable projects or ideas for implementation is critical. MAUT is gaining notice as a valuable tool in this context, offering a systematic and objective approach to evaluate and prioritize potential innovation projects based on multiple criteria, highlighting its ability to enhance decision-making processes, foster strategic thinking, and drive sustainable development.

Innovation thus plays a pivotal role in mining to unlocking new opportunities to improve productivity and secure sustainable development. However, the selection of innovation projects or ideas is often subjective and filled with uncertainties. MAUT provides a robust framework to evaluate and compare various innovation options objectively.

MAUT, if applied within a strategic innovation planning process, could enable mining companies to make more informed and rational decisions by considering multiple attributes or criteria simultaneously. By defining and quantifying these criteria, such as safety, productivity, cost, environmental impact, and social acceptability, MAUT allows for a comprehensive evaluation of innovation alternatives. This holistic approach helps decision makers gain a deeper understanding of the potential benefits and drawbacks of each option.

One of the key advantages of MAUT is its ability to quantify and assign weights to different criteria. By utilizing data-driven approaches, experts and stakeholders can assign relative importance to each attribute based on their specific goals and preferences. This process helps establish a clear hierarchy of criteria, ensuring that decision making is objective and transparent.

Considering that the mining industry operates in a complex and uncertain environment, MAUT provides a mechanism to incorporate uncertainties and risks into the decision-making process. Through sensitivity analysis and scenario modeling, decision makers can assess the robustness of different innovation options under various scenarios. This capability allows for more robust and resilient decision making, reducing the likelihood of unforeseen consequences.

MAUT could assist the mining industry by providing strategic thinking by aligning innovation selection criteria with the company's long-term objectives and core values. By including sustainability considerations, such as environmental impact and social acceptability, in the decision-making process, mining companies can prioritize innovations that align with their commitment to responsible mining practices. MAUT helps integrate sustainability as a core criterion, thereby driving the industry toward a more sustainable and socially responsible future.

Innovation projects in the mining industry often involve multiple stakeholders, including employees, communities, regulatory bodies, and investors. MAUT facilitates stakeholder engagement by providing a structured framework for collecting and incorporating their preferences and perspectives into the decision-making process. This inclusive approach fosters transparency, accountability, and collaboration, leading to greater acceptance and support for the selected innovations.

Mining companies face resource constraints, in terms of both financial and human capital. MAUT could assist in optimizing resource allocation by identifying innovations that provide the highest utility or value for the allocated resources. By considering the trade-offs between different attributes and their associated costs, MAUT helps maximize the return on investment and improve the overall efficiency of resource allocation.

MAUT encourages a culture of continuous improvement and learning within the mining industry. By regularly reviewing and updating the selection criteria and weights, companies can adapt to the changing market conditions, technological advancements, and evolving stakeholder expectations. MAUT offers a flexible framework that allows for iterative decision making, ensuring that the innovation selection process remains relevant and effective over time.

Chapter 8

The future of mining

Smart

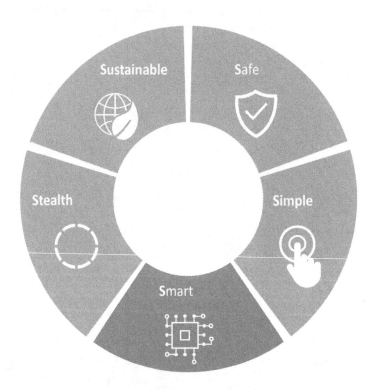

After safety and simplification, the third ultimate operational condition in the mine of the future is maximum productivity. In this case, it is possible to get the highest level of productivity using smart and autonomous technology.

Today, technology transformation is advancing so rapidly that it is difficult for any industry, including mining, to anticipate a potential loss of competitiveness due to the emergence of new technologies, always with the threatening possibility of suddenly being displaced by competition due to technology disruption. Worldwide, all industry is indeed facing a tremendous technological transformation: The Fourth Industrial Revolution (IR4).

IR4 consists of several advanced technology trends such as automation, big data analytics, AI, machine learning, VR and AR, 3-D printing, and several other technologies (Galantucci et al., 2019). IR4 is already here and will radically transform how the industry thinks and operates – a transformation so radical that in just one generation will completely require a new set of human skills and competencies that currently do not exist.

DOI: 10.1201/9781032622699-8

Among several advanced technologies evolving from the ongoing transformation that will indeed transform industry forever, two of them are specifically important to achieve a smart operational condition in the mine of the future: *Smart automation* and *smart data analytics*; these two technology drivers must, of course, be supported on a robust digital infrastructure driven by a technology implementation strategy, now called *Digital Transformation.*

Smart automation, based heavily on real-time sensorial technology, advanced robotics, and, of course, AI, is the next evolutionary step in automation in mining. Future automated mining equipment will move beyond operating autonomously but will do so by adapting to variable conditions typically present in mining. Smart automated equipment will be able to dynamically communicate with other smart autonomous equipment working in a collaborative system. This new collaborative autonomous technology will result in incredibly smart and efficient systems that will require minimum human intervention. In mining, this is good news; currently, mining faces new operational challenges in extracting minerals under very difficult settings, most of the time under hazardous conditions for humans.

Smart data analytics is the next step in the development of data analysis algorithms, which will not just sort, organize, and filter data, but also analyze data in different forms, bring real-time operational deductions, and provide answers and solutions to, sometimes, unseen problems. Smart data analytics will intelligently find potential problems even before they occur and provide possible solutions.

Following is a short description of the possible technologies that can be developed or implemented to promote and achieve smarter operational conditions in mining in the near future.

Table 8.1 List of technologies related to achieving a smart operational condition in mining.

Real-time data communication networks
• Bluetooth low energy (BLE)
Internet of things (IoT)
Big data analytics
Blockchain technology
Electric battery equipment technology
Smart automation
• Autonomous equipment
• Autonomous drilling
• AI
• Machine vision
• Cognitive computing
• Self-healing machines
• Embedded sensors
• Machine learning
• Real-time sensorial technology
• Advanced robotics
• Cyber-physical production systems
• Integrated automation
• Collaborative robotics (Cobots)
• Ventilation on demand (VOD)
VR and AR, and the Metaverse
Digital twin mine
Smart ore sorting
3D printing
Remote integrated operations center
Ultra-high-precision GPS
Drone technology

Real-time data communication networks

Real-time data communication networks are critical for mining operations, as they enable teleoperation, automation, data analytics, systems optimization, and more.

A real-time data communication network is the backbone of equipment teleoperation technology, allowing mining operators to control mining equipment from remote locations, reducing the risk of injury and increasing efficiency. By using cameras and sensors, operators can remotely control mining equipment and vehicles, making it possible to work in hazardous environments.

Real-time data communication networks are also crucial for automation in mining operations. Automated systems require real-time data feedback to ensure that they are functioning correctly. The network provides the necessary data for automated processes to function smoothly, such as monitoring equipment performance and detecting faults.

A real-time data communication network also provides valuable data for analysis, such as production rates, equipment usage, and environmental factors. By analyzing this data, mining companies can identify areas for improvement, optimize processes, and increase efficiency.

Real-time data communication networks enable mining companies to optimize their operations continually and in real time. By monitoring equipment performance, tracking material flow, and analyzing production rates, mining companies can identify areas for improvement and make real-time adjustments to increase efficiency.

Real-time data communication networks also play a critical role in ensuring worker safety. By providing real-time monitoring of equipment and environmental conditions, mining companies can identify potential hazards and take immediate action to prevent accidents.

Thus, real-time data communication networks are essential and will be even more critical for mining operations in the future by providing critical data for teleoperation, automation, data analytics, optimization, and safety, and enabling mining companies to increase efficiency, reduce costs, and improve safety.

Bluetooth low energy (BLE)

BLE technology is a wireless communication protocol designed for short-range communication among devices with low power consumption requirements. It is a subset of the classic Bluetooth technology and is specifically optimized for applications that prioritize energy efficiency and cost effectiveness.

BLE is designed to operate on small, coin cell batteries for extended periods, ranging from months to several years. It achieves this by utilizing an optimized power management scheme where devices spend most of their time in a low-power sleep state and wake up only when necessary. BLE operates in the 2.4 GHz ISM band and typically provides a communication range of up to 100 meters. However, the effective range can vary depending on factors such as interference and obstacles. BLE has a lower data transfer rate compared to classic Bluetooth. It supports data rates up to 1 Mbps, which is suitable for the applications that involve transmitting small packets of data at regular intervals.

BLE technology can play a significant role in improving communications and real-time monitoring and tracking of miners and equipment in underground mining operations, discussed as follows.

1. Miner tracking and safety: BLE-enabled tags or wearable devices can be attached to miners, allowing their real-time location tracking within the mining area. These tags can transmit signals to strategically placed BLE access points throughout the mine. By monitoring the signal, the system can determine the exact location of each miner. This capability enhances safety by enabling quick response in emergencies, such as cave-ins or accidents, as well as by providing evacuation guidance.
2. Asset tracking: BLE technology can be used to track and monitor equipment, vehicles, and tools within the mining operation. Each asset can be equipped with a BLE tag or sensor, transmitting data regarding its location, status, and condition. By deploying BLE beacons throughout the mine, the system can collect this data in real time, enabling efficient asset management and preventive maintenance.
3. Communication and messaging: BLE can facilitate reliable and low-power communication between miners and supervisors. Miners can wear BLE-enabled devices, such as headsets or smartwatches, allowing them to send and receive messages, alerts, and instructions. This communication channel can enhance coordination, provide safety updates, and allow for immediate response in critical situations.
4. Work condition monitoring: BLE sensors can be deployed to monitor underground conditions such as temperature, humidity, gas levels, and air quality within the mine. These sensors can transmit data wirelessly to a central monitoring system, which can then analyze the information in real time. This enables early detection of hazardous conditions and potential risks, allowing for proactive measures to ensure miners' safety.
5. Integration with existing systems: BLE technology can be integrated with other mining systems and technologies such as mine planning and management software, mine ventilation systems, and emergency response systems. By incorporating BLE data into these systems, operators can gain a comprehensive understanding of the mine's status, optimize operations, and make informed decisions based on real-time information.
6. Data logging and analytics: BLE-enabled devices can capture and log data from various sensors, providing a valuable source of information for analysis and optimization. This data can be used to identify patterns, predict maintenance needs, optimize workflows, and improve overall efficiency and productivity in the mining operation.

Internet of things (IoT)

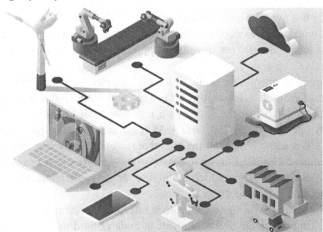

In recent years, IoT technology has become one of the most important technologies to interconnect mechanical devices, equipment, and practically any object through a 'network of things', thus being able to interact as machine to machine (M2M) to achieve wireless automation and control.

Some current uses and benefits of IoT, in both surface and underground mining, consist of tracking and receiving data from the entire mining operation, including drills, loaders, trucks, and other equipment. Its benefits are to improve overall efficiency, where all assets are networked and are able to make decisions in microseconds. Another added benefit is the capability to generate scenarios based on the acquired data to optimize production or operations.

Besides, IoT value consists in using equipment data to determine optimal ways to use the machines and make immediate diagnoses of equipment problems to reduce downtime, plus the efficient use of energy and the ability to better control possible operational interruptions.

Big data analytics

Big data analytics is a set of technologies and mathematical developments designed to analyze, store, and organize large data sets of information to find trends or patterns that can be translated into useful information. Once properly compiled, analyzed, and evaluated, big data in combination with automation can accelerate mining operational tasks such as drilling, excavation, processing, and separation of minerals.

This technology also plays a vital role in the safety, security, and reliability of mining operations. Indicators of potential mining accidents can be identified using big data analytics by processing behavioral and biological activity such as movement, location, heart rate, body temperature, and by collecting data from the working environment such as gas concentration, temperature, suspended particles.

Big data is also a prime tool to help analyze equipment operational reliability and optimize the spare parts use. Thus, predictive maintenance can be achieved by using big data to reduce disruptions. In addition, big data can assist optimizing the procurement of future services, significantly reducing overall procurement costs.

With automated and optimized transportation, big data can help improve transportation efficiency, reduce overall operational costs, and identify areas for cost improvement.

Blockchain technology

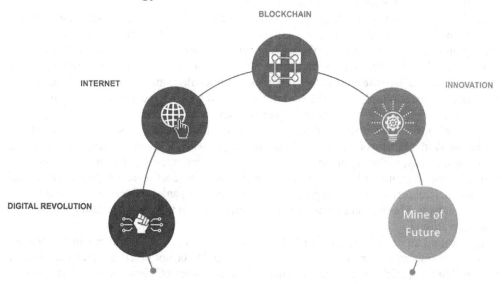

Blockchain technology has the potential to revolutionize many industries, including the mining industry. The mining industry faces numerous challenges in terms of ensuring transparency and traceability in the production and supply chain of minerals.

Blockchain is a technology that allows the traceability of data and information and is shared between transactional agents. In mining, blockchain technology can assist tracing any object or element within the mine supply chain, including the material being mined from the mine face to the mill and to the market. By using blockchain technology in mining, it will be possible to identify, at any time, the transactional status of any digital or physical asset being handled throughout the value cycle within the mining operation and throughout every level within the overall corporation.

Transactional events within a blockchain process can include, among many others, location, time, state, quality, grade, value, and any other spatial, operational, or quality indicator, making the information completely traceable and indisputable. Blockchain generates a reliable, hard-to-hack, shared, and immutable transaction record, which facilitates the process of recording transactions, as well as tracking assets, which can be tracked and traded in a network of blockchains within the operation, reducing risk and costs.

By leveraging blockchain technology, the mining industry can address these challenges by monitoring transactional events within the production and supply chain, ultimately improving efficiency and trust.

Blockchain technology can be a fundamental tool in managing mining supply chains because of its ability to continuously track transactional events of project financials, production data, and environmental and social factors, from the time an operation begins until mine closure and reclamation.

Blockchain technology offers a decentralized, immutable, and transparent system of record keeping. This means that all parties involved in the mining industry, from mining companies to buyers, can access and verify the same data without intermediaries. This eliminates the need for a centralized authority, making the system more efficient, secure, and transparent. By leveraging blockchain technology, mining companies can track the entire lifecycle of minerals, from the moment they are extracted to the moment they are sold.

Besides, blockchain technology offers smart contracts. These are self-executing contracts that automatically execute when predetermined conditions are met. For the mining industry, this means that contracts between mining companies and buyers can be programmed to execute when minerals are mined, transported, and sold. This can reduce transaction costs, minimize human error, and increase trust between parties. Additionally, smart contracts can ensure that mining companies comply with the environmental and social standards, as they can be programmed to execute only if certain conditions are met.

In addition, blockchain technology can ensure that the mining industry meets ethical and social standards. Many consumers are becoming more aware of the ethical implications of the products they buy, including the minerals used in their electronics. Blockchain technology can provide a transparent and immutable record of the entire production and supply chain, ensuring that minerals are ethically sourced and that mining companies adhere to labor and human rights standards. This can ultimately improve the reputation of the mining industry and increase consumer trust.

By creating an immutable record of all transactions, blockchain technology makes it more difficult for fraudsters to manipulate records or engage in corrupt practices. This can increase trust within the mining industry and ultimately lead to a more efficient and sustainable supply chain.

Blockchain technology, thus, offers numerous benefits to the mining industry. By providing a decentralized, transparent, and immutable system of record keeping, mining companies can track the entire lifecycle of minerals, reduce transaction costs, and increase trust between parties. Additionally, smart contracts can ensure that contracts are executed automatically when predetermined conditions are met, ensuring compliance with ethical and social standards. Ultimately, blockchain technology can lead to a more efficient, sustainable, and trustworthy mining industry.

Electric battery equipment technology

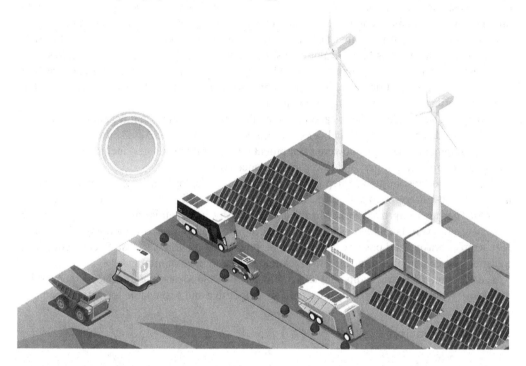

Mining operations can be challenging as they go deeper, due to the harsh and demanding environments they are often located in. To overcome these challenges, the industry has increasingly turned to technology to improve efficiency, productivity, and safety. One such technology is electric equipment, which includes battery-based, hydrogen-electric, and other types of electric-powered machinery. In this part, we will explore the importance of using electric equipment technology in mining and the benefits it provides.

One of the primary advantages of using electric equipment in mining is increased reliability. Traditional diesel-powered equipment is prone to breakdowns, which can cause delays and reduce productivity. Electric equipment, on the other hand, has fewer moving parts and is less prone to mechanical failure. Additionally, electric equipment can be easily monitored and maintained, reducing the risk of unexpected downtime. This increased reliability translates into improved operational efficiency and reduced maintenance costs.

Another significant advantage of electric equipment is its increased productivity. Electric machinery can operate with greater precision than traditional diesel-powered machinery. This increased accuracy can result in higher throughput and faster cycle times, increasing overall productivity. Additionally, electric equipment can be equipped with advanced sensors and monitoring systems, providing real-time data that can be used to optimize operations and improve efficiency.

The use of electric equipment also has significant cost-saving benefits for mining operations. Electric machinery is generally more energy efficient than diesel-powered machinery, resulting in lower fuel costs. Additionally, electric equipment requires less maintenance, reducing maintenance costs and downtime. The reduced costs associated with electric equipment can help mining companies remain competitive in an industry where profit margins can be thin.

Improved ventilation is another benefit of electric equipment in mining. Traditional diesel-powered equipment emits harmful pollutants, including carbon dioxide, nitrogen oxides, and particulate matter. These pollutants can build up in underground mining operations, creating hazardous working conditions. Electric equipment produces no harmful emissions, reducing the need for ventilation and improving air quality for miners.

Hydrogen-electric technology is another type of electric equipment that is gaining traction in the mining industry. Hydrogen fuel cells can provide a clean and reliable source of power for mining equipment. Hydrogen-powered equipment has several advantages, including zero emissions, longer operating times, and lower maintenance costs.

The next and most important step in mining is the mass deployment of zero-emission equipment. Electric-battery-operated vehicles and equipment not only save on ventilation and cooling costs but also improve worker health and safety by being emission free and generating less noise and overall machine vibration. Advances in reliability, electric autonomy, electrically regenerative brakes, and power capacity, among others, will continue to lead to cost reductions in battery-based equipment, making the technology more viable in mining.

The adoption of electric equipment will continue to drive innovation and efficiency in the mining industry, helping mining companies remain competitive and sustainable for years to come.

Smart automation

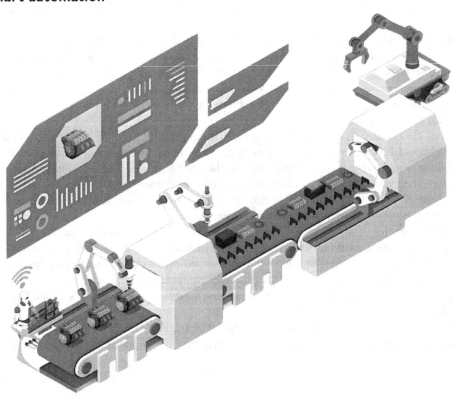

In recent years, smart automation has emerged as a solution to address several challenges the mining industry is facing today and most likely will continue to face in the future.

Smart automation involves the use of advanced technologies such as AI, machine vision, cognitive computing, embedded sensors, machine learning, real-time sensorial technology, and collaborative robotics to automate mining operations. The integration of these technologies into mining operations can help increase efficiency, improve safety, and reduce operational costs.

One of the most significant applications of smart automation in mining is the use of AI and machine learning algorithms to automate various processes such as drilling, blasting, and hauling. These algorithms can analyze data from sensors and machines, predict equipment failures, and optimize operations to improve efficiency and reduce downtime.

Machine vision technology can also be used in mining operations to enhance safety and productivity. By using cameras and sensors, machine vision systems can detect equipment malfunctions, monitor the environment for potential hazards, and track workers' movements to ensure their safety. Cognitive computing is another technology that can be used in mining to improve decision making and problem solving. By analyzing data from various sources such as sensors, equipment, and geological data, cognitive computing systems can provide insights and recommendations to operators, enabling them to make informed decisions. Embedded sensors are also vital components of smart automation in mining. These sensors can be attached to equipment to monitor various parameters such as temperature, pressure, and vibration, allowing operators to detect potential issues before they lead to equipment failure. Real-time sensorial technology is another crucial component of smart automation in mining. This technology can be used to monitor the environment and detect potential hazards such as gas leaks or unstable ground conditions, enabling operators to take corrective action before an accident occurs. Collaborative robotics can also play a significant role in mining operations. These robots can be used to perform tasks that are too dangerous or difficult for human workers, such as operating in underground mines or in hazardous environments. Collaborative robots can work alongside human operators, improving productivity and safety.

As technology continues to advance, the use of smart automation in mining operations is expected to grow, further enhancing the efficiency and safety of the industry. Below is a summary of key technologies essential to the evolution of smart automation and robotics in mining. Following is a set of key technologies currently enabling smart automation in mining.

Autonomous equipment

The mining industry has always been an area of significant economic activity, but with the advent of technology, the way mining operations are carried out is rapidly evolving. The introduction of autonomous vehicles and equipment into the mining sector is a revolutionary development that has transformed the mining industry into a smart industry.

The use of autonomous vehicles and equipment in mining operations has the potential to bring numerous benefits to the industry by directly improving safety. Mining is a hazardous activity, and the use of autonomous vehicles and equipment eliminates the need for humans to perform risky operations. It minimizes the risk of human injury and improves safety at the workplace.

The use of driverless vehicles and autonomous equipment in mining also improves efficiency. By using autonomous equipment, mining operations can be carried out more efficiently, accurately, and quickly. Autonomous vehicles can operate 24/7, improving the productivity of mining operations.

Also, the smart automation of equipment reduces operating costs. Autonomous vehicles and equipment do not require rest or sleep, and they can operate for longer periods without the need for maintenance or repair. They also improve fuel efficiency and reduce energy consumption, resulting in significant cost savings for mining companies.

With the rapid deployment of autonomous and connected vehicle technology, by using wireless vehicle-to-vehicle (V2V) communication, any mining vehicle can exchange accurate information (e.g., position, speed, acceleration, jolt, and transit information) and timely adjust the routes, course of travel, and speed according to the anticipation and interactive participation of traffic conditions underground as well as on the surface.

The integration of autonomous vehicles and equipment into mining operations is also crucial for the future of sustainable mining. The use of autonomous vehicles and equipment can help reduce this impact. For instance, autonomous equipment can be programmed to optimize the use of energy, reducing the carbon footprint of mining operations.

Finally, the use of autonomous vehicles and equipment in mining operations allows for better data collection and analysis. The data collected by these vehicles and equipment can be used to improve the overall efficiency of mining operations, optimize production processes, and identify areas for improvement.

Mining companies that embrace these technologies will not only gain a competitive advantage but also contribute to the overall progress of the mining industry.

Autonomous drilling

Rock drilling is an essential process in mining; it plays a very important role in productivity, cost, and efficiency. Different geological conditions require different mining methods, and different mining methods require different types of rock drilling. During the past decade, the mining industry has been incorporating advanced drilling technology into their operations; drilling equipment has been automated, integrating communication technologies, sensors, AI, VR, control equipment, and more. These technologies allow the generation of accurate, reliable data which is the base of precise operational decisions during the drilling process. Autonomous drilling equipment can be operated and monitored from a remote command center. Holes are drilled in a pre-established order at the locations and depths loaded into the system. The drill is precisely guided to the appropriate hole coordinates for precise hole location. When the desired depth is reached, the drill string is retracted and then pushed into the next hole. The path-planning capability allows for the optimized movement of multiple drills. Drills are guided to the hole coordinates within the drill rig using a planning system; this allows autonomous navigation to other drill rigs, or safe zones, during blasting operations. Autonomous drilling allows mining production to be kept at an optimal level and improves mining efficiency and economic benefits. In this way, environmentally friendly, safe, and efficient mining can be achieved.

Artificial intelligence

AI is based on complex mathematical models that transform and develop machines capable of performing a given task without requiring any specific instruction. The final objective in the field of AI is, for machines, to reach the level of human intelligence. In this context, AI and machine learning can be applied from the beginning of mining to the end of the mine life cycle, or from prospecting to production, closure, and mine remediation.

AI is being used in mining operations to improve safety, increase efficiency, and reduce costs. AI can be used to automate processes, such as data collection and analysis, to identify patterns and trends in data, and to optimize production processes. AI can also be used to monitor equipment and personnel, detect anomalies, and provide predictive maintenance. AI can also be used to improve decision making, such as in the selection of mining sites and in the optimization of mining operations.

Vision-based AI is being developed to introduce decision-making capability into haulage trucks to react according to the working environment. AI will allow a better classification of the terrain and the environment, generating better perception technology. In the field of mineral processing, AI along with machine learning and big data analytics will be applied to the flotation process, with minerals of any composition. Modern algorithms, such as neural hybrids, deep learning, and convolution networks with exceptional pattern recognition and knowledge extraction capabilities will be used in mining processes, such as drilling, blasting, prospecting, and mineral exploration, for intelligently enhanced operational automation. All this development and implementation of smart, intelligence-based technology will be an essential technology driver in order to reach the mine of the future operating as a nearly fully automated operation, thus resulting in an extremely efficient, safe, and sustainable operation.

Machine vision

Machine vision is an automated vision system that can be implemented in mining equipment working in underground and surface mines. Machine vision, or visual inspection system, consists of three or more of the following elements: Image camera(s), vision algorithms that process and interpret the image, sensors, vision processing systems, and communication systems.

The idea that machines could see and act as a 'human operator' is not new; machine vision can process images that are not constrained to use the visible light spectrum (standard video images), and it can also process images produced by heat, infrared, or even radar sensors, or other sources of information that could be converted into a referential image. Machine vision is capable of processing 3D images and automatic learning, and addressing the needs and challenges typically present in underground mining. Machine vision working in real-time can be very useful in monitoring underground conditions, early detection of leaks, detection of equipment overheating, fire prevention, early detection of mechanical failure, remote visual validation of critical events, among others. With the advancement of AI and the convergence among optical detection, thermal detection, and video analysis, it is feasible to generate several layers of analyses and degrees of decision making, independent of human intervention. For example, in visual monitoring, or environmental-risk-mitigation applications (the risks associated with leaks or spills is of the highest priority in any mining operation), this technology automates the detection and notification of events based on pre-established parameters; thus allowing personnel working from a remote-control center to quickly react to system warnings in seconds, visually validate events from their remote location, and make immediate decisions.

Cognitive computing

The growing adoption of digitalization and automation has driven the development of cognitive computing. Cognitive computing is the creation of self-learning systems that use data analytics, pattern recognition, and natural language processing to emulate how the human brain works. The purpose of cognitive computing is to create computer systems that can solve complicated problems without constant human supervision. Cognitive computing incorporates technologies such as natural language processing, machine learning, automated reasoning, and information retrieval, which are used in the translation of unstructured data to perceive, infer, and predict the best solution, at various stages of the mining value chain, for exploration, mine development, mineral processing, and logistics.

Self-healing machines

Self-healing machine technology is based on arrays of embedded sensors, as well as on AI and machine learning technologies. Mining equipment using self-healing technology can be designed to recognize variations in machine parts performance and correct them before they fail causing problems that require downtime and repair. This technology has the potential to save time and money and frees up employees, who would normally monitor equipment and perform maintenance to work on higher-level tasks.

Embedded sensors

Embedded sensor technology is used to detect changes in equipment in real time, and is able to share physical and logical sensorial information among local and remote smart systems and equipment. Physical sensorial information can include temperature, light, pressure, sound, and motion. Logical sensorial data includes the presence or absence of an electronically traceable unit, location, or activity. Sensors that are essential in the mining environment may include, among others, those that can detect chemicals, gas, radiation, leakage, flow, movement, acceleration, temperature, acoustics, ambient light, optics, electrical pulse, force, and pressure. As the IoT evolves, and as electronic miniaturization advances, embedded sensors will become standard in future mining equipment, giving way to a whole new age of connectivity, systems integration, and data ecosystem in the intelligent mine of the future.

Machine learning

Machine learning is an application of AI integrated by a wide variety of algorithms and models that learn, in an iterative way, from the processing of active data in order to improve, describe, and predict results, providing, in this case, to the mining systems the capacity to learn from the operation and improve the tasks automatically. Machine learning can detect patterns from operational data that are useful for reducing costs and optimizing resources.

Machine learning is a crucial component of smart automation technology in mining operations. Smart automation technology involves using advanced sensors, data analytics, and machine learning algorithms to automate various aspects of mining operations such as drilling, hauling, and processing.

Here are some specific ways in which machine learning is important in smart automation technology for mining.

1. Predictive maintenance: Machine learning algorithms can analyze data from the sensors on the mining equipment to identify the patterns that indicate potential equipment failures. This allows mining companies to perform maintenance before a breakdown occurs, reducing downtime and increasing efficiency.
2. Autonomous vehicles: Autonomous vehicles are becoming increasingly common in mining operations, and machine learning is a critical component of their operation. Machine learning algorithms can analyze data from the sensors on these vehicles to help them navigate the mine safely and efficiently.
3. Ore sorting: Machine learning algorithms can analyze data from the sensors on ore-sorting equipment to identify and separate different types of ores. This allows mining companies to extract valuable minerals more efficiently and reduce waste.
4. Environmental monitoring: Machine learning algorithms can analyze data from environmental sensors to identify potential hazards and prevent environmental damage.

Machine learning is thus essential for making the smart automation technology in mining operations more efficient, safe, and sustainable. It enables mining companies to extract minerals more efficiently, reduce downtime, and improve worker safety, all while minimizing environmental impact.

Real-time sensorial technology

The advancement of embedded real-time sensory technology during the past decade has been steadily maturing. New mobile, sensory, and data network technologies are available, serving the mining industry in order to adapt to the new operational challenges required for the new digital age.

These technologies cover the following applications: Operations through AR, assisted operation, and automation of machinery; traceability of ore, machinery, and personnel; access control and identity through biometric tags and indicators; and more.

Advanced robotics

Mining companies worldwide have been modernizing their operations since several years now, developing and implementing heavily automated systems within their operations; the next step in automation though is advanced robotics. Advance robotics means the successful application of AI, machine learning, into automated system in order to create a self-organizing and collaborative automated system within the mine operation. Advanced robotics are machines, or systems, capable of accepting high-level commands; they are flexible, intelligent, and autonomous. The potential benefits of advanced robotics are abundant, as mine activities involve strenuous working conditions and are potentially dangerous for the miner. In other words, mining robots in the future will perform more complex tasks, making mining operations safer for humans and keeping operators away from work hazards produced by dust, noise, gas emissions; proximity to heavy machinery; and many others. The smart mine of the future will be based on massive connectivity of systems and devices through digital networks, with a high degree of smart automation using mine robots that act largely, or partially, autonomously, that physically interact with people or their environment, and that can modify their behavior based on sensorial data.

Cyber-physical production systems

They are systems of autonomous and cooperative elements that are connected to each other, at all levels of the mining processes, through machinery and equipment to the production and logistics networks, improving the decision-making processes in real time, and responding to both unforeseen conditions and changes in evolution over time.

The cyber-physical production systems are usually connected to each other, and also to remote data storage and management services. A cyber-physical production system can be formed by sensors with connectivity, by devices linked to the IoT that are capable of generating data and sending it, or by robots that can perform different tasks. The characteristics that drive the rise of cyber-physical production systems in the mine of the future are focused on increasing the processing capacity of the devices, reducing their size, improving connectivity, interoperability between different systems of the mining processes, the increasing use of information storage systems, and the application of AI systems.

Integrated automation

Integrated automation is an open system architecture that guarantees the smooth interaction of all automation components, software involved, systems, and services. Its development is strongly conditioned by standardization efforts in control engineering, computer engineering, and information technologies; and it adheres to global standards and uniform interfaces from field level to enterprise management level.

The integrated automation of the entire mining operation has intelligent devices and equipment, enabled for autonomous configuration, efficient operation, and self-diagnosis; and is equipped with software that offers total transparency and functional performance in real time. The intelligent integration covers all relevant technologies from advanced sensors and real-time process control to modeling, visualization, and optimization. It manages the entire mining support mechanism, facilitating visibility of resources across the mine, real-time production performance, market conditions, available ore types, and optimal response actions for critical mining asset conditions.

Collaborative robotics (Cobots)

The combination of automation, AI, and robotics are indeed contributing to a rapid acceleration of an ongoing transition in the mining industry to a smarter, more efficient, and sustainable production model. A cobot is a robot designed to physically interact with other machines and humans in a shared workspace. At the industrial level, there is increasing adoption of cobot technology, as companies including mining realize the immense potential of automation. Robots and humans have their own strengths and limitations; the advanced safety features of cobots allow employees to work alongside robots safely without producing blind spots or obstacles at the workplace. It is proved now that cobot technology has the potential to increase mine operator satisfaction and reduce the risk of workplace accidents. Cobot tech also increases productivity and when applicable improves the overall quality of a given product.

Ventilation on demand (VOD)

Ventilation is an essential aspect of underground mining operations as it provides fresh air to workers and removes harmful gases and dust generated during mining activities. Traditionally, ventilation systems are designed to provide a constant flow of air regardless of the actual mining conditions, leading to unnecessary energy consumption and reduced air quality. The implementation of VOD has emerged as a promising solution to address these challenges. Some advantages of implementing VOD in underground mining are as follows.

Implementing VOD in an underground mining operation will reduce energy consumption and costs associated with the operation. A traditional ventilation system operates at a constant rate irrespective of the actual need for air in any particular zone or area within the mine. This results in the use of more power than necessary, leading to higher energy and power costs. VOD systems, on the other hand, operate based on real-time data gathered from the sensors placed around the mine, which measure air quality and determine the actual need for ventilation. As a result, VOD systems use less energy and lower the associated energy costs.

VOD systems provide a safer working environment for mine workers. Underground mining can expose workers to harmful gases such as carbon monoxide produced by blasting, nitrogen oxides, and diesel exhaust fumes from underground mining equipment. VOD systems help mitigate these risks by ensuring that workers are always breathing clean air. In addition, VOD systems can respond quickly to changes in the environment, such as rock falls or fires, by increasing ventilation rates to ensure worker safety.

Besides the above, VOD systems provide more precise control over air quality. The traditional ventilation systems provide a constant flow of air, which can sometimes lead to low-quality air in certain areas of the mine. VOD systems, on the other hand, can provide varying amounts of air to different parts of the mine based on the level of mining activity and the number of workers present. This ensures that air quality is always maintained at safe levels.

VOD systems may also assist in reducing the need for expensive capital investments. Traditional ventilation systems require significant capital investments to construct and maintain. However, VOD systems require less initial capital investment as they use existing infrastructure and are typically retrofitted onto existing ventilation systems. Additionally, VOD systems require less maintenance and can reduce overall maintenance costs.

At last, VOD systems can improve productivity and mine efficiency. By optimizing ventilation rates based on real-time data, VOD systems can help reduce downtime associated with ventilation system maintenance and improve overall mine productivity. Additionally, VOD systems can help reduce the time required to ventilate an area of the mine, allowing for quicker access to working face.

Overall, implementing VOD systems in underground mining operations has numerous advantages. These systems can reduce energy consumption and costs, provide a safer working environment for mine workers, improve air quality, reduce capital investments, and improve overall mine productivity. As such, the adoption of VOD systems should be considered as a way to improve the sustainability and efficiency of underground mining operations.

VR, AR, and the Metaverse

VR, AR, and the Metaverse are innovative technologies that have the potential to revolutionize the mining industry. These technologies allow mining companies to visualize and simulate complex systems and processes, improving decision making, safety, and efficiency in mining operations. Several advantages of using VR, AR, and the Metaverse in mining operations are discussed as follows.

VR is a technology that creates a simulated environment that the user can interact with using a headset and controllers. This technology can be used in the mining industry to provide training simulations for miners, allowing them to practice procedures and techniques in a safe, virtual environment before performing them in real-life situations. This not only improves safety but also reduces the risk of equipment damage and downtime due to mistakes made by inexperienced operators. Additionally, VR can be used to create virtual tours of mining sites, allowing investors and stakeholders to explore the mine without physically being present there, reducing travel costs and increasing accessibility.

AR is a technology that overlays digital information onto the user's physical environment, typically using a smartphone or tablet, and in the near future through AR goggles. AR can be used in mining operations to provide real-time data on equipment performance, allowing maintenance and repair teams to quickly identify issues and take corrective action. AR can also be used to provide miners with contextual information on their surroundings, such as safety guidelines and warning signs, improving safety and reducing the risk of accidents.

The Metaverse is a virtual world where users can interact with each other and digital objects in real time, using avatars to represent themselves. This technology has the potential to revolutionize the way mining operations are managed and conducted, allowing for remote control and monitoring of equipment, and reducing the need for workers to be physically present on site. Additionally, the Metaverse can be used to create collaborative workspaces, allowing teams from different locations to work together on projects in real time, improving efficiency and reducing the need for travel.

The advantages of using VR, AR, and the Metaverse in mining operations are numerous. These technologies can improve safety by providing workers with virtual training simulations and real-time information on their surroundings. This reduces the risk of accidents and equipment damage caused by inexperienced operators or workers not being aware of potential hazards. Also, these technologies can improve efficiency by allowing for remote monitoring and control of equipment, reducing the need for workers and miners to be physically present on site. This improves travel logistics, reduces costs, and increases productivity and safety of mining sites, particularly in remote or hazardous locations.

Besides, these technologies can improve decision making by providing stakeholders with virtual tours of mining sites and real-time data on equipment performance. This improves transparency and collaboration, allowing for better communication and decision making between mining operational and logistics teams. At last, these technologies can reduce the environmental impact of mining operations by allowing for more precise and efficient use of resources. For example, virtual simulations can be used to optimize drilling and blasting techniques, reducing waste and minimizing their impact on the surrounding environment.

The advantages of using VR, AR, and the Metaverse in mining operations are numerous and significant. These technologies have the potential to significantly improve the way mining operations are managed and conducted, improving safety, efficiency, and sustainability. As the mining industry continues to face challenges such as resource depletion, environmental concerns, and the need for more efficient operations, these technologies will become increasingly important in ensuring a sustainable future of mining and the supply of critical minerals.

Digital twin mine

Leveraging algorithms and AI to process value chain data is increasingly important to provide real-time decision support and future projections. In addition to the technologies mentioned, there is the emerging digital twin technology, which has the ability to capture the status of intelligent production, transportation, and processing systems in real time and predict system failures through optimization. A digital twin is a digital replica, a carbon copy, of a physical entity. It allows a seamless transfer of data by connecting the physical and virtual worlds. In its simplest form, the digital twin is a virtual simulator of mining operation – a copy in the digital world where management can manipulate a large number of variables, over a given period of time, to see how changes will affect mining processes.

Digital twin technology has the potential to transform the mining industry by enabling mining companies to create virtual models of their physical assets and processes. Some ways in which digital twin technology could help the mining industry are discussed as follows.

By creating digital twins of mining equipment, mining companies can implement advanced predictive maintenance to monitor the performance of equipment in real time and predict when maintenance is required. This can help reduce downtime and maintenance costs.

Digital twins can be used to optimize mining operations to simulate different scenarios and identify areas for improvement in mining processes. This can help optimize the use of resources and increase productivity.

By creating digital twins of mining sites, companies can identify potential safety hazards and take corrective action before accidents occur. Also, digital twins can be used to train new employees on mining processes and equipment in a safe and controlled environment, reducing the risk of accidents and increasing efficiency.

Digital twins can be used in environmental monitoring processes to monitor the environmental impact of mining operations and identify the ways to reduce that impact, as well as in supply chain management to track the movement of materials and equipment throughout the mining process, improving supply chain management and reducing the risk of delays.

Digital twin technology has the potential to transform the mining industry by improving efficiency, safety, and sustainability while reducing costs.

Smart ore sorting

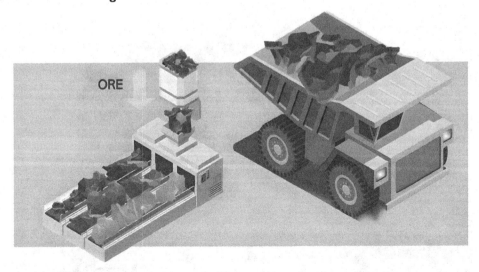

One application of intelligent automation in the mining industry is sensor-based ore classification. This is also known as ore classification, automated classification, electronic classification, particle classification, or optical classification.

Ore sorting is done by analyzing the transported bulk materials in real time and diverting the unwanted material from the ore feed. Today, there are intelligent sorting machines that can classify the extracted material from the mine based on the measurement of some chemical and/or physical property in order to differentiate the valuable material from the waste. Sensor-based technologies include radiometric (RM), X-ray transmission (XRT), X-ray fluorescence (XRF), couple-charged device (CCD) color camera, photometric (PM), near-infrared spectrometry (NIR), infrared (IR), microwave (MW), and electromagnetic (EM). In addition, there are other technologies under development such as hyperspectral sensors visible in the near-infrared (VNIR) and short-wave infrared (SWIR) which, in combination with automatic learning routines, improve the automated sorting process. Ore sorting has the potential to improve overall material handling process in any mining operation including improving the ore quality being fed to the processing plant, reducing tailings outputs, reducing transportation costs per unit of production; minimizing water consumption; improving energy and fuel consumption; reducing waste costs; thus improving the overall productivity and efficiency of the operation.

3D printing

3D printing or manufacturing is a form of rapid prototyping using additive manufacturing technology, in which a 3D object of any kind is created, or made, by accumulating successive layers of materials in order to forma a 3D object. The advantages of 3D manufacturing are production flexibility and availability, reducing implementation cost and time. 3D manufacturing is today a feasible and more reliable manufacturing technology, thanks to the rapid evolution of 3D scanning technologies such as computerized tomography or photogrammetry. Mining companies operating in remote locations will extensively use 3D printing to manufacture on-site spare parts for equipment and machinery, minimizing delivery times and downtime. Similarly, a custom-designed 3D printed drill bit can be produced in one day, compared to conventional practices that take even months to complete. With digitally stored spare parts, storage and inventory costs can be reduced, and the lengthy and costly process of transporting parts to remote sites can be eliminated.

Remote integrated operations centers

The mining industry is gradually moving toward mine control through remote operation centers, by connecting information and communication technologies with real-time operational technologies, in order to exchange data throughout the mining operations and the supply chain. Remote operation centers are centralized and connected control rooms for mines and metallurgical plants, providing an external environment for personnel to collaborate in operations, without having to be present at the site itself. Due to improved network connectivity and bandwidth capacity, these control rooms can be located anywhere in the world. Remote operation centers provide real-time control of operations in remote locations where there is minimal infrastructure. By providing video streams and other digital tools, they enable employees to monitor and control multiple aspects of operations simultaneously. Remote operation centers also integrate multiple data sources to support decision making based on real-time mine condition, or metal processing conditions, allowing for monitoring and coordination of multiple services. Remote operation centers provide advanced operational awareness without needing highly qualified specialists; remote centers also allow for better reaction to tactical production problems and emergency situations, more efficient and reliable operations, better production performance, and better strategic collaborative decisions. Remote control of logistics is another source of significant savings. In addition, remote operation centers also reduce the land footprint of mining sites.

Summary list of technologies related to achieve a smart operational condition in mining.

Ultra-high-precision GPS

High-precision GPS technology is extremely important nowadays for accurate mapping and surveying in surface mining operations. Ultra-high-precision GPS technology will further improve mine tracking and surveying systems which will be able to measure in real time at a sub-centimeter precision. This level of precision will be used to create very accurate maps and surveys of mining sites. New generation of GPS location technology will help mining companies plan their operations in more detail, more efficiently, better identify potential hazards, and optimize the use of resources.

High-precision GPS technology, being more reliable and precise in the future, will exhaustively be used to monitor and track personnel, material handling systems, and equipment. This technology will become a working commodity for mining companies in order to operate locally or/and remotely their mines in any region around the globe.

High-precision GPS technology will continue to be an important safety technology in mining operations. By accurately tracking the movement of vehicles and equipment, mining companies can ensure that they are operating in safe areas and avoid collisions and other accidents.

High-precision GPS technology can help mining companies operate more efficiently by significantly reducing the time and resources needed for mapping, surveying, and tracking, resulting in important cost savings and increased productivity. High-precision GPS technology can also be used to monitor and manage the impact of mining operations on the environment. By accurately tracking the movement of vehicles and equipment, mining companies can avoid environmentally sensitive areas and minimize their impact on the surrounding ecosystem.

Drone technology

The use of drone technology has revolutionized the mining industry by providing a range of benefits such as improved monitoring, safety, logistics, operations, security, and 3D mapping LiDAR for surveying. Drones are small unmanned aerial vehicles that are equipped with high-resolution cameras, LiDAR, and other sensors that can capture data from an elevated plan view.

One of the main benefits of drone technology in mining is improved monitoring. Drones can be used to monitor mining operations in real time, providing managers with up-to-date information on the progress of mining activities, the location of equipment, and the status of mine sites. This allows mine managers to quickly respond to any issues that arise, improving efficiency and reducing downtime.

In terms of safety, drones can be used to conduct inspections of mining equipment and structures, reducing the need for mining workers to perform dangerous tasks. Drones can also be used to monitor the movement of vehicles and equipment, ensuring that they are operating in safe areas, and in combination with GPS and radar technology to avoid potential collisions.

Logistics is another area where drone technology can be beneficial. Drones can be used to transport equipment and supplies to remote mining sites, reducing the need for ground-based transportation and improving the efficiency of logistics operations.

In terms of security, drones can be used to monitor mining sites and detect any unauthorized access or activity. This can help prevent theft and other security breaches, improving the safety and security of mining operations.

One of the most exciting applications of drone technology in mining is 3D mapping LiDAR for surveying. LiDAR is a technology that uses lasers to create detailed 3D maps of mining sites, providing accurate and up-to-date information on the topography of the land, the location of resources, and the progress of both underground and surface mining operations.

The use of drone technology in surface and underground mining is quickly transforming the mining industry by providing a range of benefits such as improved monitoring, safety, logistics, operations, security, and 3D mapping LiDAR for surveying. As drone technology continues to advance, we can expect to see even more applications of this technology in mining, improving efficiency and safety.

Chapter 9

The future of mining
Stealth

Stealth mining is the fourth optimum operational condition in mining. During the lifespan of a mining operation, large footprints are created (Sinha et al., 2019). By executing an effective innovation strategy, mining footprints can be significantly reduced. Innovative mining methods and new mining equipment will assist in attaining a minimal footprint or stealth operational condition. Footprint minimization will inevitably steer mining operations toward the use of underground mining methods. Most likely, the mine of the future will be underground, heavily automated, and remotely teleoperated.

Compared to the high production levels seen in surface mining operations, underground mining is typically constrained to lower production rates due to the confined nature of the operation.

DOI: 10.1201/9781032622699-9

The key to high production rates in underground mining is to achieve in some degree the three operational conditions mentioned earlier – safe, simplified, and smart operations – plus the development and implementation of new underground extractive and processing methods and strategies aimed to minimize material handling all underground, from blasting to extraction, recovery, and processing of the mineral.

The evolution of the mine of the future into becoming a *stealth* operation consists of three tech mining phases: (1) The first phase is the total automation of transporting the ore material from the underground face or stope to a processing plant located on the mine surface. (2) The second phase is to handle the blasting, processing, and recovery of the ore material, all underground. (3) The third phase is using advanced in situ recovery methods that will leach the mineral content directly from the host rock without the need to use traditional blasting and processing methods.

Advanced highly monitored underground caving mining is also a promising approach to be able to mine at large production rates. New caving-monitoring technology is being developed today that will help track in real time the caving flow within the caved stope as the material is being extracted (Chitombo, 2018).

Surface mining will not completely disappear in the near future. It will still be used by the minerals industry for several years; however, due to operational flexibility and lower costs, surface mining methods of the future will need to drastically evolve from digging massive open pits into digging temporary open trenches, mined by fully autonomous equipment; in this case, open trenches, once fully depleted, will be filled in and environmentally resorted to drastically reduce mining footprint.

Following are the examples of innovative methods and technology trends being used today to achieve a stealth operational condition.

Table 9.1 List of technologies related to achieving a stealth operational condition in mining.

Zero emissions
• **Emissions and noise minimization by electric systems**
In-place mining
• **In-line recovery methods**
• **In-mine recovery methods**
• **In-situ recovery methods**
• **Biohydrometallurgy**
Advanced cave mining and real-time underground monitoring

Zero emissions

The minerals extractive industry has traditionally been associated with environmental impact, including greenhouse gas (GHG) emissions, noise pollution, wastewater discharge, and tailings disposal. These environmental impacts have adverse consequences on both environment and human health. It is, thus, imperative that mining companies adopt zero-emission strategies to mitigate the environmental impacts of their operations.

One of the primary concerns associated with mining activities is GHG emissions. The mining industry is a major contributor to global GHG emissions, with most emissions generated through the combustion of fossil fuels in mining operations. These emissions have severe implications on the environment and human health.

To mitigate the impact of GHG emissions, mining companies must adopt a zero-emissions strategy. This strategy should focus on reducing energy consumption, using renewable energy sources, and implementing carbon capture and storage technologies to sequester CO_2. These strategies can significantly reduce the carbon footprint of mining operations, thereby reducing their impact on the environment and human health.

Another important aspect of zero-emission strategies in mining is the reduction of noise pollution. Mining operations are typically associated with significant noise pollution, which can have adverse effects on wildlife and human health. Noise pollution can disrupt wildlife behavior, cause hearing damage in humans, and contribute to stress-related health issues. Therefore, mining companies also need to consider adopting noise-reduction strategies, including the use of low-noise equipment, such as electric equipment and noise barriers, and operational practices that minimize noise levels.

Wastewater discharge is another significant environmental concern associated with mining activities. Mining operations typically generate wastewater, which can contain toxic chemicals and heavy metals that can contaminate water sources and harm aquatic life. To mitigate the impact of wastewater discharge, mining companies must adopt wastewater treatment technologies that can remove contaminants and ensure that the water is safe for discharge into the environment.

Tailings disposal is another environmental concern associated with mining activities. Tailings are the waste materials generated during the mining process, and they often contain toxic chemicals and heavy metals. Improper disposal of tailings can lead to soil and water contamination, which can harm the environment and human health. Therefore, mining companies must adopt tailings management strategies and new technologies that minimize the generation and containment of tailings.

The adoption of zero-emission strategies using new waste management technology in mining is crucial for mitigating the environmental impacts of mining activities. These strategies should focus on reducing GHG emissions, minimizing noise pollution, implementing wastewater treatment technologies, and adopting tailings management practices that prioritize safe disposal and containment. By adopting these strategies, mining companies can significantly reduce their environmental footprint promoting stealthier operations.

Emissions and noise minimization by electric systems

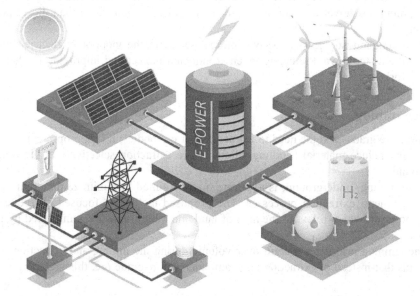

The use of electric equipment in mining operations can improve potential environmental impact as well as miners' safety and health. As mentioned before, replacing fossil fuel (diesel) with electric equipment can significantly reduce noise and emissions, and improve mine ventilation in underground mining.

Noise is a significant hazard in mining operations, as it can lead to hearing loss and other health problems. The use of electric equipment in mining can significantly reduce noise levels. Electric equipment is generally quieter than diesel-powered equipment, reducing noise pollution and improving the work environment for miners. This reduction in noise levels can also improve communication and safety in mining operations, as it allows for better communication between workers and reduces the risk of accidents due to noisy equipment.

Diesel engines used in mining equipment produce harmful emissions such as particulate matter, carbon monoxide, and nitrogen oxides. These emissions can cause respiratory problems, among other health-related issues. In contrast, electric equipment produces zero emissions at the point of use, reducing the environmental impact of mining operations. By reducing the emissions from mining equipment, the mining industry can improve air pollution, promoting the wellbeing of communities near mining operations.

Ventilation is a critical aspect in underground mining operations as it is essential to maintain a safe working environment for miners. Poor ventilation can lead to the buildup of harmful gases, such as methane, carbon monoxide, and hydrogen sulfide, which can cause respiratory problems, explosions, and fires. The use of electric equipment in mining significantly improves ventilation in underground mining by reducing the amount of heat generated by equipment. Electric equipment produces less heat than diesel-powered equipment, reducing the need for ventilation to remove heat from underground mines. This reduction in the amount of heat generated can improve the effectiveness of ventilation systems and reduce the energy required to maintain safe working conditions.

In addition to the benefits of reduced noise, emissions, and improved ventilation, the use of electric equipment in mining can also reduce operating costs. Electric equipment is generally more energy-efficient than diesel-powered equipment, reducing fuel consumption and operating costs. The use of electric equipment can also reduce maintenance costs, as electric equipment has fewer moving parts and requires less maintenance than diesel-powered equipment.

As the mining industry seeks to become more sustainable, the adoption of electric equipment is an important step toward reducing its environmental impact and improving the health and safety of workers.

In-place mining

With lower ore grades and the need to extract deeper ore bodies, it is appropriate to use novel methods that reduce material handling, allowing ore extraction selectively at the mine face underground.

In-place mining, underground, is a mining method aimed to minimize the environmental impact of mineral extraction processes. The proposed method is focused on reducing the mining footprint, which refers to the area of land and natural resources impacted by mining activities.

In-place mining prioritizes margin over volume during the mining and extraction process. This means that instead of extracting large amounts of low-grade ore, this method focuses on extracting higher-grade ore. This allows mining companies to reduce their overall mining foot-print while maintaining profitability.

Another benefit of in-place mining is that most of the material remains underground. This is in contrast to the traditional mining methods where a large amount of material is removed from the site and transported to the processing facilities. With in-place mining, underground, the ore is extracted directly from the mine, and only the valuable minerals are processed underground (in-place) or brought to the surface for processing.

In addition, in-place mining allows for smaller underground openings. This is because this method does not require large openings (surface mining) or tunnels to be excavated, which can take up a significant amount of space. Instead, smaller, more targeted openings are created to access the ore, which reduces the amount of excavation required.

Another advantage of in-place mining is reduced waste disposal. With traditional mining methods, a significant amount of waste rock is generated, which must be disposed of. This can be a major source of environmental contamination, as waste rock can contain harmful chemicals and heavy metals. With in-place mining, the amount of waste rock generated is significantly reduced, as only the valuable minerals are extracted.

Additionally, in-place mining requires smaller surface infrastructure. This means that there is less need for buildings, roads, and other facilities that can take up a significant amount of space. This can reduce the overall mining footprint and minimize its impact on the surrounding environment.

Moreover, this mining approach is aimed to keep most of the mining activities underground, resulting in less need for heavy machinery and equipment on the surface. This reduces the visual impact of mining activities and helps preserve the natural landscape.

In-place mining, underground, can reduce energy and water use. This is because less material is excavated, transported, and processed, which requires less energy and water. Additionally, in-place mining has the potential to significantly reduce the need for waste tailings storage, which can further reduce energy and water use, thus assisting to create a more sustainable mining industry.

In-line recovery methods

This method proposes using innovative production strategies aiming for higher quality rather than quantity. The in-line recovery concept is aimed for recovering of minerals integrating selective mining, thus upgrading the quality of the ore, allowing just the rich-content material to be pre-concentrated at the mine face, to be then transported to the processing facility located either underground or on the surface.

This method is based on automated or semi-automated equipment located at the extraction point to selectively extract and sort high-grade material, significantly reducing material handling costs. In-line recovery is driven by the operational value of minimal material handling and processing, which also promotes a minimal surface footprint, reducing tailings and encouraging automation, with a significant overall operational cost reduction.

In-line mining technology is particularly important at depth in deep mining operations where overall mining costs increase considerably, and human safety becomes a significant issue.

In-mine recovery methods

In-mine recovery methods favor a minimum or zero rock material movement to the surface and may involve fracturing ore in place, employing more advanced technology for mechanical fracturing in order to liberate the high-grade ore from the host rocks. This technology promotes mechanical excavation technologies to allow effective fracturing and generates a conditioned mineral block.

This approach can be combined with chemical and/or biological tools to leach the ore under unsaturated conditions. In-mine recovery methods are aimed to treat and process the broken ore in underground facilities using innovative metallurgical recovery processes situated all underground.

The operational objective of this method is to be able to extract and pump the high-grade ore in concentrate slurry form to the surface for further refining, smelting, or transport to market. The aim of in-mine recovery methods is to promote in-mine processing underground, thus further minimizing the overall mine footprint. Mine recovery methods will enable the cost-effective extraction of minerals that are currently not viable, as well as the reduction of energy consumption by reducing activities associated with transport, milling, waste disposal, and surface infrastructure.

In situ recovery methods

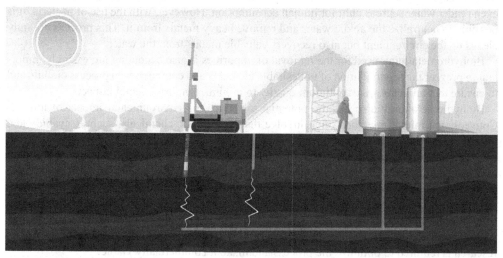

In situ recovery is an advanced form of in-mine recovery, which involves the processing of the in situ ore by in situ leaching with no host-rock material being blasted nor moved. The mineral concentrate resulting from the in situ leaching process is then pumped to the surface to be further enriched, all with minimal ore material handling.

When feasible, it is a simple, safe, and cost-effective method. In situ recovery methods, such biohydrometallurgy, are not widely used by the mineral industry today because the geological conditions must be right. There are few places in the world that meet the following conditions: The ore body must be naturally broken, fractured, and highly permeable; the target ore must be soluble in the proper fluid, typically a weak acid; and the ore deposit must be below the phreatic level to allow fluid movement through the ore body. Some of the added advantages of in situ recovery methods in a mining operation include minimum noise, minimum dust or GHG emissions, and minimum disturbance. In addition, on-site recovery methods reduce capital and operating costs while creating a safer environment for mine workers.

Biohydrometallurgy

Biohydrometallurgy is an emerging field of study that is poised to revolutionize the way minerals are extracted from ores. Biohydrometallurgy refers to the use of microorganisms, particularly bacteria, and archaea, in the extraction of metals from ores. This process is environmentally friendly and has the potential to significantly reduce the environmental impact of mining. Future applications of biohydrometallurgy in the minerals industry bring some advantages due to its ability to extract metals from ores that are currently considered uneconomical. For example, some copper deposits contain low-grade ore that cannot be processed profitably using conventional methods. However, with the use of biohydrometallurgy, it is possible to extract copper from these ores by utilizing bacteria that can leach copper ions from the ore. This could potentially open vast new resources of copper and other metals that were previously considered uneconomical.

Another application of biohydrometallurgy is in the treatment of acid mine drainage. Acid mine drainage is a significant environmental issue that is caused by the release of acidic water from abandoned mines. This acidic water can cause severe damage to the environment and can even render water sources unfit for human consumption. However, with the use of bacteria, it is possible to neutralize the acidic water and remove heavy metals from it. This process not only cleans up the environment but also recovers valuable metals from the water.

Biohydrometallurgy used for the removal of impurities is also becoming increasingly important to prevent the accumulation of undesirable elements and compounds in process circuits and to ensure that effluents are of optimum quality to minimize environmental impacts.

Biohydrometallurgy also has the potential to reduce the environmental impact of mining. Conventional mining methods often involve the use of toxic chemicals, such as cyanide and mercury, to extract metals from ores. These chemicals can be harmful to the environment and can pose a risk to human health. In contrast, biohydrometallurgy uses bacteria that are naturally occurring and pose no threat to the environment. Additionally, the use of bacteria can significantly reduce the amount of waste produced during the mining process.

One of the challenges facing biohydrometallurgy is the optimization of the process. The use of bacteria in the extraction of metals is a complex process that is influenced by many factors such as temperature, pH, and the presence of other minerals. Therefore, a significant amount of research is required to optimize the process and make it commercially viable.

This technology has been applied commercially in the extraction of base metals from low-grade sulfide ores, and in the pretreatment of refractory sulfide ores containing gold. Biohydrometallurgy uses the activity of microorganisms in aqueous extractive metallurgy to extract value from minerals, concentrates, and wastes. It is considered a promising and environmentally friendly option for moving to a circular economy, especially for low-grade and complex minerals.

The future of biohydrometallurgy in the minerals industry is promising. Its ability to extract metals from uneconomical ores, treat acid mine drainage, and reduce the environmental impact of mining makes it an attractive alternative to conventional methods. However, further research is required to optimize the process and make it commercially viable.

Advanced cave mining and real-time underground monitoring

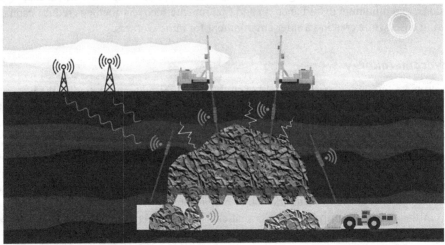

The development of new and improved cave mining methods aims to ensure that underground mining remains a viable, optimized, and automated operation in the future, in order to meet the essential requirements associated with safety, profitability, and efficiency.

To this end, advanced techniques for cave mining have incorporated several technological advances, including autonomous mining systems, geotechnical monitoring, and technologies for tunnel development, tunnel boring machines, micro-seismic monitoring, remote monitoring for cave management, hydro-fracturing for cave propagation, jaw crushers, hydro-fracturing for rock mass pre-conditioning, and electric loaders.

In the future, the elaboration of basic open communication protocols and the future development of specific automation platforms and their use are intelligent solutions that will optimize the viability, real-time 3D monitoring, cost reduction, profitability, and production when applying caving methods in an underground mining operation.

Chapter 10

The future of mining

Sustainable

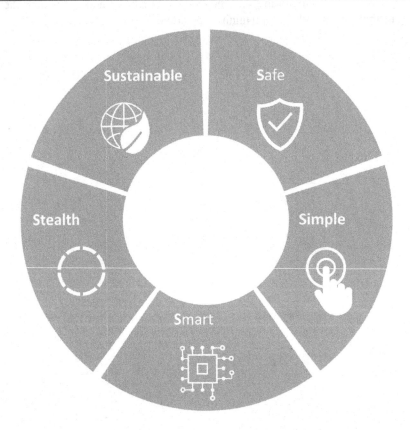

Sustainability is the fifth condition in a mining operation; as described earlier, the mine of the future most likely will be evolving into underground systems in order to minimize the footprint of the overall operation. The sustainability of the operation along with the sustainability of the attached ecosystem is in this case the fifth ultimate operational condition of the mine of the future. An optimal operational condition is achieved when the overall mining system is self-sustained in terms of the resources it needs to operate. The mine of the future will use only renewable energy and will reduce water use to a bare minimum, using highly advanced recycling waste and water systems using continuous sensorial systems aiming to have a mining operation with zero discharge.

A full sustainable condition in the mine of the future will consist in a fully harmonic operation with its surroundings, promoting social and economic development in the region. Sustainability

DOI: 10.1201/9781032622699-10

in mining must promote economic development during and after operations and must fully embrace the total preservation and enhancement of the local ecosystem.

In this case, the innovation model for the mine of the future promotes sustainability through technology innovation, promoting sustainability practices through the application of new technology and also promotes regional economic development, maintaining and improving local ecosystems. A sustainable condition in the mine of the future also considers the first condition described in this model – total safety.

Following are examples of methods and technologies related to achieve a sustainable operational condition in mining.

Table 10.1 List of methods and technologies to achieve a sustainable operational condition in mining.

Renewable energy technology
• **Solar energy**
• **Wind energy**
• **Geothermal energy**
• **Hydropower**
• **Ocean energy**
• **Bioenergy**
Recyclable water and waterless systems
• **Hydraulic dewatered stacking**
• **Coarse particle recovery**
Advanced alternative clean power source systems
Environmental, social, and governance (ESG) through technology
Wellbeing through tech innovation

Renewable energy technology

Renewable energy sources can play a significant role in helping mining operations become sustainable. By reducing reliance on non-renewable energy sources such as coal, oil, and natural gas, renewable energy sources can help mining companies reduce their environmental impact, lower their operating costs, and enhance their social license to operate.

There are several renewable energy sources that can be employed in mining operations, including solar energy, wind energy, geothermal energy, hydropower, ocean energy, and bioenergy. Each of these energy sources has unique characteristics that make it suitable for specific applications in the mining industry.

Solar energy

Solar energy is the conversion of sunlight into electrical energy using photovoltaic cells. The use of solar energy in mining operations can significantly reduce the energy costs associated with running the mine. The installation of solar panels can also help reduce the carbon footprint of mining operations by reducing the reliance on fossil fuels. In addition, the use of solar energy can enhance the social license to operate of mining companies by reducing the impact of their operations on local communities and the environment.

Wind energy

It is the conversion of wind energy into electrical energy using wind turbines. The use of wind energy in mining operations can provide a reliable source of electricity, especially in remote locations where access to the electricity grid is limited. The use of wind energy can also help reduce the carbon footprint of mining operations by reducing the reliance on fossil fuels. In addition, the use of wind energy can enhance the social license to operate of mining companies by reducing the impact of their operations on local communities and the environment.

Geothermal energy

Geothermal energy is the conversion of heat energy from the earth's interior into electrical energy. The use of geothermal energy in mining operations can provide a reliable source of electricity, especially in locations with high geothermal potential. The use of geothermal energy can also help reduce the carbon footprint of mining operations by reducing the reliance on fossil fuels. In addition, the use of geothermal energy can enhance the social license to operate of mining companies by reducing the impact of their operations on local communities and the environment.

Hydropower

Hydropower is the conversion of the energy of falling water into electrical energy. The use of hydropower in mining operations can provide a reliable source of electricity, especially in locations with high water potential. The use of hydropower can also help reduce the carbon footprint of mining operations by reducing the reliance on fossil fuels. In addition, the use of hydropower can enhance the social license to operate of mining companies by reducing the impact of their operations on local communities and the environment.

Ocean energy

Ocean energy is the conversion of the energy of ocean currents, waves, and tides into electrical energy. The use of ocean energy in mining operations can provide a reliable source of electricity, especially in coastal locations with high ocean energy potential. The use of ocean energy can also help reduce the carbon footprint of mining operations by reducing the reliance on fossil fuels. In addition, the use of ocean energy can enhance the social license to operate of mining companies by reducing the impact of their operations on local communities and the environment.

Bioenergy

Bioenergy is the conversion of biomass into electrical energy. The use of bioenergy in mining operations can provide a reliable source of electricity, especially in locations with high biomass potential. The use of bioenergy can also help reduce the carbon footprint of mining operations by reducing the reliance on fossil fuels. In addition, the use of bioenergy can enhance the social license to operate of mining companies by reducing the impact of their operations on local communities and the environment.

Thus, renewable energy sources can play a significant role in helping mining operations to become sustainable.

Recyclable water and waterless systems

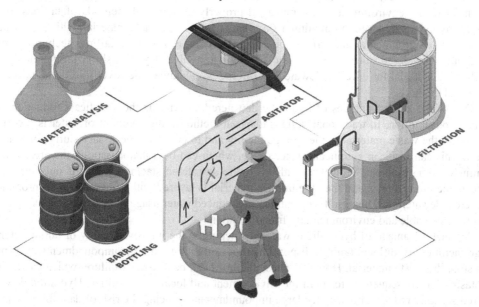

Mining operations are resource dependent on the geographical location of the ore to be extracted, very often with limited options to mitigate potential water scarcity and water quality in the region. Water sustainability and water stress in mining operations pose complex challenges that require transformative technology solutions to reduce water consumption and promote water reuse by embracing strategies aimed to achieve a zero-discharge operation using smart treatment and recycling technologies. Effective water reuse technology will be essential in the mine of the future, along with proper management of water supply sources, to reduce overall water consumption. Thus, the search for new methods to optimize water consumption must focus on developing technologies and recycling solutions to comply with current and future regulations. The use of advanced water treatment technologies and alternative sources of supply has created an opportunity for alternative use of seawater in some regions. Desalination processes for water use in mining operations will have even more relevance in the mine of the future.

Mining operations can potentially reduce the use of fresh water by 50%, by increasing the level of water recycling. Current and future technologies aimed to eliminate the use of fresh water in mining processes, and eventually achieving a near-waterless mine, will promote the development of new ways to separate and transport waste, manage evaporation, encourage the use of dry tailings, and develop advanced non-aqueous mineral processing. Currently, there are several promising technologies in the area of pre-concentration: Paste filling and dry stacking of waste; reprocessing of existing tailings (with important environmental benefits as well as economic returns); deployment of a closed-loop system aimed at minimizing water losses by direct water recycling and reuse; and real-time automated control systems of all water streams for water saving.

Hydraulic dewatered stacking

Mining operations produce large quantities of waste material, or tailings, which can be harmful to the environment if not managed properly. Traditional methods of tailings disposal involve using large impoundments or dams to store the waste material, which can lead to several environmental and safety concerns. However, new mining tailings technology of hydraulic dewatered stacking or dry stacking, which involves layering slurry tailing material with sand material for efficient dewatering, offers several advantages over traditional tailings disposal methods.

One of the key advantages of hydraulic dewatered stacking is the significant reduction in water consumption. In traditional tailings disposal methods, large amount of water is used to transport the waste material to the storage facility and to create the necessary slurry for transport. This can result in significant water usage, which can be a major issue in the areas with limited water resources. However, with hydraulic dewatered stacking, water is only used to create the initial slurry, and then the tailings material is layered with sand material to promote efficient dewatering. This results in significantly reduced water usage of up to 80%, making it a more sustainable and environmentally friendly option.

Another advantage of hydraulic dewatered stacking is the improved safety of tailings storage facilities. Traditional tailings disposal methods often rely on large impoundments or dams to store the waste material. However, these facilities can be at risk of failure, which can have catastrophic consequences for both the environment and local communities. Hydraulic dewatered stacking reduces the need for large impoundments, reducing the risk of dam failures and improving overall safety.

Additionally, hydraulic dewatered stacking can improve tailings management practices. With traditional tailings disposal methods, it can be difficult to control and manage the waste material, which can result in environmental damage and increased risk of accidents. However, with hydraulic dewatered stacking, the tailings material is layered with sand material, which promotes efficient dewatering and allows for better control over the disposal process. This can lead to improved tailings management practices and a reduced environmental impact.

As the mining industry continues to seek ways to reduce its environmental impact and improve sustainability, the adoption of dry stacking tailing technology will likely become increasingly widespread.

Overall, these are some of the key advantages of using dry stacking technology in mining.

1. Reduced water consumption: The use of hydraulic dewatering or dry stacking reduces the amount of water needed to dispose of tailings, resulting in significant water savings.
2. Increased safety: Traditional tailings disposal methods can lead to the risk of dam failures, which can be catastrophic for both the environment and local communities. Hydraulic dewatering or dry stacking eliminates this risk.
3. Reduced environmental impact: Traditional tailings disposal methods often result in discharge of large amounts of contaminated water into nearby rivers and streams. Hydraulic dewatering or dry stacking greatly reduces this environmental impact.
4. Improved land use: The use of hydraulic dewatering or dry stacking allows for the rehabilitation of previously used tailings storage facilities, freeing up land for other uses.
5. Improved tailings management: The use of hydraulic dewatering or dry stacking allows for better control over tailings disposal, which can lead to improved management practices and reduced environmental impacts.

Coarse particle recovery

The mining industry is faced with the challenge of keeping productivity by reducing its environmental impact and improving its sustainability. One way to achieve this goal is through the use of coarse particle recovery during the comminution and recovery metallurgical process. This technology involves the process and recovery of minerals based on coarse particles from the ore stream, which can reduce water and power use, and improve dry stacking of tailings.

One of the key advantages of coarse particle recovery is the reduction in water usage. In traditional comminution and metallurgical processes, a large amount of water is used to recover the valuable minerals from the ore stream. However, with coarse particle recovery, the valuable minerals can be recovered from the ore stream using less water. As a result, coarse particle recovery can significantly reduce water usage, making it a more sustainable and environmentally friendly option.

Another advantage of coarse particle recovery is its ability to improve dry stacking of tailings. New dry stacking methods involve the layering of tailings material with sand material for efficient dewatering. However, fine particles can make it difficult to achieve effective dewatering, which can result in increased water usage. Coarse particle recovery can help reduce fine particles volumes in the tailings stream, making it easier to achieve effective dewatering and improve the efficiency of the dry stacking process.

The use of coarse particle recovery during the comminution, and separation recovery process offers several advantages over traditional methods. It can reduce water usage and improve dry stacking of tailings by maintaining good recoveries in mining operations. The adoption of coarse particle recovery technology will likely become increasingly widespread in mining during the following years.

Advanced alternative clean power source systems

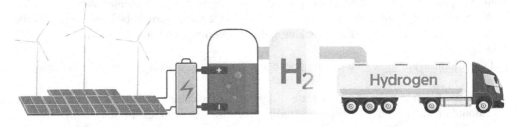

Today, concerns about climate change and carbon footprint are driving the mining industry to optimize their mining operation to comply with global and state regulations, promoting the use of advanced energy-efficient mining technology to reduce GHG emissions.

One of the main contributors to these environmental challenges is the use of fossil fuels as the primary source of energy in mining operations. As the world moves toward a more sustainable future, alternative clean fuels have emerged as viable options for powering mining systems. As such, the use of advanced clean power sources in mining, such as solar power, wind power, geothermal, green hydrogen, and hydropower, is becoming increasingly important.

The use of solar power in mining can significantly reduce GHG emissions. Solar power is a renewable energy source that does not produce GHG emissions during operation. This is particularly important for mining operations that are located in remote areas where access to traditional power sources is limited. Solar power can provide a reliable source of energy that can reduce the need for diesel generators, which are a significant source of emissions.

Wind power can also be an important source of clean energy in mining. Wind turbines can be used to generate electricity to power mining equipment and operations. Like solar power, wind power is a renewable energy source that does not produce GHG emissions during operation. Additionally, wind power can be particularly useful for mining operations that are located in wind favorable areas.

Hydrogen is another alternative clean fuel that has gained significant attention in recent years. Hydrogen is a versatile fuel that can be produced from a range of sources. However, not all hydrogen is created equal, and different production methods can result in different environmental impacts. Gray hydrogen is produced from fossil fuels and is the most carbon-intensive form of hydrogen production. Blue hydrogen is produced from fossil fuels but includes carbon capture and storage (CCS) technology to reduce carbon emissions. Green hydrogen is produced from renewable energy sources and is the cleanest form of hydrogen production.

One of the main advantages of using hydrogen as fuel in mining systems is its versatility. Hydrogen can be used in a range of mining operations, from powering heavy machinery to providing energy for processing and refining operations. Additionally, hydrogen fuel cells are highly efficient, with a conversion efficiency of up to 60% compared to 20–30% for internal combustion engines. This efficiency can lead to significant cost savings over time, making hydrogen an attractive option for mining companies looking to reduce costs and improve operational efficiency.

Green hydrogen can be used as an alternative to traditional diesel fuel. It is produced by electrolysis using renewable energy sources, such as wind and solar power, and can be used to power mining equipment and operations.

Hydropower is another clean energy source that can be used in mining. Hydropower uses the energy of moving water to generate electricity. This can be particularly useful for mining

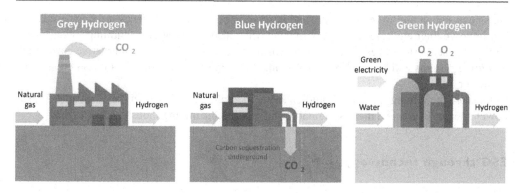

Figure 10.1 Types of hydrogen according to generation source.
Source: Nieto (2019b)

operations that are located near water sources such as rivers or dams. Like solar and wind power, hydropower is a renewable energy source that does not produce GHG emissions during operation.

One of the main advantages of using alternative clean fuels in mining systems is the reduction of carbon emissions. Fossil fuels such as coal, oil, and gas release a large amount of carbon dioxide into the atmosphere when burned, contributing to climate change. In contrast, alternative clean fuels such as solar, wind, hydroelectric, and geothermal energy do not produce any carbon emissions during the generation of electricity. By using these clean energy sources, mining operations can significantly reduce their carbon footprint and contribute to the global efforts to mitigate climate change.

Another advantage of using alternative clean fuels in mining systems is the reduction of air and water pollution. Fossil fuels release a wide range of pollutants, including sulfur dioxide, nitrogen oxides, and particulate matter, which can have significant impacts on local air quality and human health. Additionally, mining operations that rely on fossil fuels can also contaminate local water sources through the discharge of wastewater and other pollutants. By using alternative clean fuels such as solar, wind, hydroelectric, and geothermal energy, mining operations can significantly reduce air and water pollution, leading to improved environmental and human health outcomes.

The use of advanced clean power sources can also help reduce costs and improve operational efficiency. Energy-related costs are a major factor in any mining operation as mining, per its operational nature, is a heavy energy-demanding activity. Renewable energy sources, such as solar and wind power, have become increasingly cost competitive in recent years, and can provide a reliable source of energy that can reduce the need for traditional power sources, such as diesel generators.

In addition, the use of advanced clean power sources can also help improve the social and economic sustainability of mining operations. By reducing the impact of mining activities on the environment and local communities, mining companies can improve their social license to operate and build stronger relationships with local stakeholders.

Thus, advanced clean power sources in mining, such as solar power, wind power, green hydrogen, and hydropower, are becoming increasingly important.

The use of alternative clean fuels such as green electricity and hydrogen in mining systems offers significant advantages over traditional fossil-fuel-based systems. By reducing carbon emissions and air and water pollution, and improving operational efficiency, mining companies can improve their environmental performance while also reducing costs and improving profitability. As the world moves toward a more sustainable future, the use of alternative clean fuels in mining systems will become increasingly important in ensuring a sustainable and prosperous future for all.

As such, the mining industry should continue to explore and invest in these advanced clean power sources to create a more sustainable and responsible mining industry.

ESG through technology

It is well known that the mining industry creates exceptional value as a developer of local communities in any country around the world. The mining industry actively works to promote environmental and climate awareness and assumes social responsibility for both employees and those who live and work in the vicinity of mining operations. However, mining operations can have negative impacts on the environment and society, making it crucial for companies to adopt sustainable practices that mitigate these effects. One way to achieve this is through the use of technology, which can assist in social responsibility and ESG initiatives.

Technology can play a crucial role in promoting social responsibility in the mining industry. For instance, the use of automation and AI can help reduce the risk of accidents, promote worker safety, and improve working conditions. Additionally, digital platforms can enable better communication between mining companies and local communities, providing a platform for engagement and collaboration on sustainable development initiatives.

Furthermore, technology can aid in improving ESG practices in the mining industry. For example, the use of drones and satellite imagery can help companies monitor and reduce their environmental impact, by identifying the areas of biodiversity and monitoring water quality. Similarly, the adoption of blockchain technology can enhance transparency and accountability, providing an immutable record of supply chains and the sourcing of materials.

The adoption of technology to enhance social responsibility and ESG practices in the mining industry is critical for improving social wellbeing. By embracing sustainable practices, mining companies can help ensure that their operations have a positive impact on the communities they operate in while also supporting the long-term viability of their businesses.

Moreover, sustainable mining practices can contribute to the economic growth and social development of local communities, by providing job opportunities and supporting local supply chains. This can help promote social wellbeing, reduce poverty, and support the United Nations' Sustainable Development Goals.

Technology can play a vital role in promoting social responsibility and ESG practices in the mining industry. By embracing sustainable practices and leveraging technological innovations, mining companies can improve social wellbeing and support long-term economic growth and development.

Today, and in the future, it will be imperative to promote local and global awareness that the mining industry through innovation and the use of state-of-the-art technology indeed creates value, promoting sustainable development and building a better future for all.

Wellbeing through tech innovation

Mining plays an essential role in the global economy, responsible for extracting valuable resources that drive industrial and technological progress. However, mining can also be associated with social and environmental risks and challenges, including safety hazards, environmental impact, and health of its workers. To address these issues, mining companies will have to embrace wellbeing as a strategic objective.

Wellbeing refers to the overall state of health and happiness of individuals and communities. In the context of the mining industry, wellbeing encompasses several dimensions, including occupational health and safety, mental health, physical health, and community health. These dimensions are closely interlinked and must be addressed comprehensively to achieve sustainable and responsible mining practices.

Mining companies must consider the broader positive implications that mining operations could bring to society via innovation and technology.

Creating real benefits for communities near mining sites will be key to successfully developing new projects. In recent years, obtaining a license to operate, from local communities, has been a challenge for the mining industry. More specifically, there is a need for co-ownership models where communities benefit from mining operations; for example, shared water infrastructure not only benefits the mining company but also serves the surrounding communities and farmers. Other means to help ensure the sustainability of a region after mine closure could include the installation of renewable energy projects on the remediated land or soil remediation processes through biohydrometallurgy.

One of the main reasons why embracing wellbeing is important in the mining industry is the need to ensure the safety of workers. Mining operations involve various hazardous activities such as drilling, blasting, and handling heavy equipment. Accidents can have severe consequences, including fatalities, injuries, and long-term health problems. By prioritizing the wellbeing of their workers, mining companies can reduce the risk of accidents and create a safer working environment.

Furthermore, investing in the physical and mental health of workers can lead to higher productivity and better job satisfaction. Healthy employees are more likely to perform better, take fewer sick days, and be more engaged in their work. This can result in increased efficiency, higher quality output, and reduced turnover rates.

Another key reason to embrace wellbeing as a strategic objective in the mining industry is its potential impact on the environment and surrounding communities. Mining operations can have negative effects on air and water quality and can cause soil erosion and biodiversity loss. By prioritizing community health and wellbeing, mining companies can mitigate these impacts and build stronger relationships with local communities.

Embracing wellbeing can help mining companies meet the growing demand for responsible and sustainable practices. Consumers, investors, and regulators are increasingly concerned about the social and environmental impact of mining operations, and companies that prioritize wellbeing are more likely to gain public trust and support.

The mining industry today and in the future must embrace wellbeing as a strategic objective to ensure the safety and health of its workers, improve productivity and job satisfaction, mitigate environmental impacts, and meet the demand for responsible and sustainable practices. By adopting a holistic approach to wellbeing, mining companies can create a culture of safety, health, and sustainability that benefits all stakeholders.

Chapter 11

Innovation roadmap to the mine of the future

Nowadays, innovation is a key strategy to maintain high levels of growth, productivity, and sustainability in the mining industry.

To stay competitive and relevant within the context of technological progress and potential disruption, defining a tech innovation model today is crucial for the mining industry of the future.

Innovation is a crucial element that businesses must consider when developing a business plan and creating a roadmap for growth. Innovation refers to the process of developing new ideas, products, services, or processes that bring value to the mining business and its customers. By introducing innovation into their operations, mining businesses will increase productivity, revenue, and growth.

One of the most significant benefits of innovation is increased productivity. Innovative solutions can help businesses streamline their operations, automate processes, and reduce waste, which can significantly boost productivity levels. This can result in cost savings, increased efficiency, and higher profits.

Innovation can also help the mining business generate more revenue by introducing new products or services into the market. By developing innovative products or services that meet the needs and wants of customers, mining companies can gain a competitive edge over their rivals and increase their market share. This can translate into higher revenues and profits, which can fuel further growth.

Innovation is essential for long-term growth. As markets and customer needs evolve, businesses must continue to innovate to remain relevant and competitive. By investing in innovation, the mining industry can stay ahead of the curve and continue to grow and expand, diversifying beyond ore commodity extraction to other added-value markets.

As discussed, the introduced mining innovation model is designed based on five key operational conditions to be achieved in any mining operation of the future. The innovation model is specifically designed for the mining industry and to be consistent with any type of mining methods including hard-rock or soft-rock, metal or nonmetal, or surface or underground.

As shown in Figure 11.3, the model defines five key operational conditions (or mining innovation drivers): (1) Achieving maximum safety, (2) simplifying systems, (3) using smart-intelligent systems, (4) designing stealth operations, and (5) following a sustainable strategy.

Human resource skills, as a sixth enabler driver, is also a critical aspect of preparing the next generation of professionals to manage new technology and innovation. In today's rapidly changing and dynamic business environment, organizations must continually innovate to remain

DOI: 10.1201/9781032622699-11

competitive. The introduction of new technologies and processes demands a skilled workforce that can adapt to new changes and utilize these advancements to drive growth.

Skills development programs offer employees the opportunity to learn new skills and keep up with the latest technological advancements in their respective fields. These programs can be formal or informal and may include training sessions, seminars, workshops, and certification programs. By investing in the development of their employees' skills, organizations can ensure that their workforce has the necessary knowledge and expertise to manage new technology and innovation effectively.

One of the benefits of skills development is that it helps foster a culture of continuous learning within an organization. When employees have access to ongoing training opportunities, they are more likely to stay engaged and motivated in their work. This leads to higher job satisfaction, lower turnover rates, and improved productivity. Furthermore, by providing employees with opportunities to acquire new skills, organizations can attract and retain top talent, which is critical in today's competitive job market.

Another benefit of skills development is that it can lead to the creation of new job opportunities. As organizations adopt new technologies and processes, they may require new roles and responsibilities to manage these changes effectively. Skills development programs can help employees acquire the skills they need to transition into new roles or to take on additional responsibilities, which can lead to career growth and advancement opportunities.

Skills development is an essential aspect of preparing the next generation of technicians and professionals to manage new technology and innovation. By investing in the development of their employees' skills, organizations can create a culture of continuous learning, attract and retain top talent, and stay competitive in a rapidly changing business environment. As technology continues to advance, it is essential that organizations prioritize skills development to ensure that their workforce has the necessary knowledge and expertise to succeed in the future.

Innovation is thus a vital component of any business plan for productivity and growth. By introducing innovative solutions, businesses can increase productivity, generate more revenue, and achieve long-term growth. Therefore, businesses that prioritize innovation are more likely to succeed in today's rapidly changing business environment.

The innovation strategy introduced in this book is designed to serve as a practical guideline to be considered in a business plan and/or roadmap for the mining industry, as shown in Figure 11.1 and Figure 11.3. An innovation strategy can also help identify new technology trends that need to be addressed today to achieve the five ultimate operational conditions, safe, stealth, simple, smart, and sustainable, to successfully operate the mine of the future.

Following are some examples of technology roadmaps currently used by the mining industry based on the 5S innovation model considering three development horizons or phases, as shown in Figure 11.1.

In Figure 11.1, vertical Y-axis of the graph indicates three development phases: Horizon 1, Horizon 2, and Horizon 3. Each horizon is defined by a time frame and a specific focus.

Horizon 1: 0–2 years. Focus on digitalization and visualization of data.
Horizon 2: 3–7 years. Focus on integration and optimization of new technologies.
Horizon 3: 7 years onward. Focus on embracing the mine of the future.

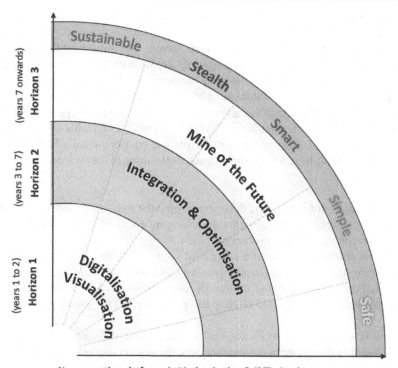

Figure 11.1 Innovation strategy, Gold Fields, 5S technology roadmap.

Source: Xenaki et al. (2023)

Horizontal X-axis of the graph indicates the innovation progress through time following the five technology drivers: Safety, simplification, smart systems, stealth systems, and sustainable systems. Each technology driver has a different level of impact on each horizon. Here's a description of the graph.

1. Safety: In Horizon 1, safety has a significant impact, as the focus is on automation, digitalization, and visualization of data. In Horizon 2, safety has a high impact, as new technologies are being integrated and optimized. In Horizon 3, safety continues to have a high impact, as the mine of the future must be safe and secure.
2. Simplification: In Horizon 1, simplification has a high impact, as the focus is on making data more accessible and easier to understand. In Horizon 2, simplification has a medium impact, as new technologies may require more complexity. In Horizon 3, simplification has a high impact again, as the mine of the future must be simple to operate and maintain.
3. Smart systems: In Horizon 1, smart systems have a high impact, as the focus is on data analysis and optimization. In Horizon 2, smart systems continue to have a high impact, as new technologies require intelligent and automated systems. In Horizon 3, smart

systems have a very high impact, as the mine of the future must be highly automated and optimized.

4. Stealth systems: In Horizon 1, stealth systems have a low impact, as the focus is on data visualization rather than security. In Horizon 2, stealth systems have a medium impact, as new technologies may require more security measures. In Horizon 3, stealth systems have a high impact, as the mine of the future must be highly secure and protected.

5. Sustainable systems: In Horizon 1, sustainable systems have a low impact, as the focus is on data visualization rather than sustainability. In Horizon 2, sustainable systems have a high impact, as new technologies must be sustainable and environmentally friendly. In Horizon 3, sustainable systems have a very high impact, as the mine of the future must be sustainable and energy efficient.

The Innovation graph in Figure 11.1, thus, shows how the five technology drivers impact the three development horizons. It highlights the changing focus of each horizon and how the impact of each technology driver evolves over time.

Figure 11.2 is another example of a successful implementation of an innovation model by the Penoles Industries, a multinational silver–zinc mining company based in Mexico City. This model is fundamentally based on the 5S model introduced in this book. The Penoles innovation model was implemented in 2020 and is the basis of their technology development platform.

Figure 11.3 is an example of technology roadmaps developed for each of the innovation 5S drivers introduced in this book. In this example, the innovation process is defined by three time horizons that can vary according to the mine company's strategic plan.

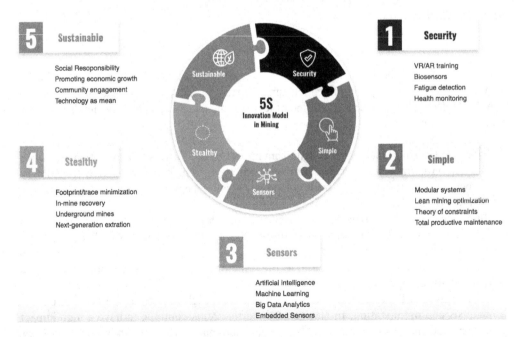

Figure 11.2 Innovation strategy, Penoles Industries.

Source: Xenaki et al. (2023)

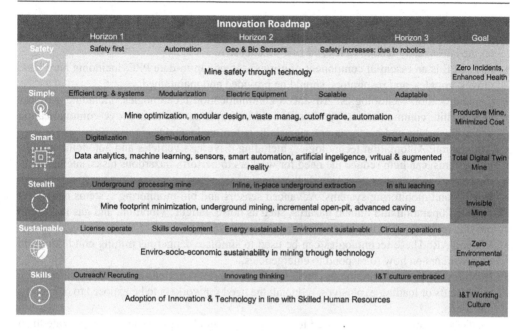

Figure 11.3 Mine of the future roadmap.

Source: Nieto (2019b)

In this example, each technology driver is described following a specific roadmap consisting of development stages, with a specific objective for each driver, as seen on the right side of the graph.

Below is a visual description of each driver including the description of a technology-skills-tech roadmap which is defined at the end of this chapter.

Safety-tech roadmap

Mining is a critical industry that provides important raw and critical materials for a wide range of sectors, but it also carries inherent risks to workers and the environment. As a result, it is essential to prioritize safety in the mining industry and invest in current and future safety technologies. The ultimate goal is to achieve zero incidents and to adopt wellbeing practices that will promote and enhance miners' health.

Safety	Safety first	Automation	Geo & Bio Sensors	Safety increases: due to robotics	
	Mine safety through technology				Zero Incidents, Enhanced Health

Figure 11.4 Safety-tech roadmap.

Here is a roadmap that considers both current and future safety technologies in mining.

- Risk assessment: The first step in any mining operation is to assess the potential risks involved. This should include a comprehensive evaluation of the geological, environmental, and operational factors that could pose a risk to the workers or the surrounding community.

- Training and education: Ensuring that workers are properly trained and educated on the potential risks and how to mitigate them is crucial. This includes providing ongoing training on the latest safety technologies and best practices.
- PPE: PPE is an essential component of mining safety. Up-to-date PPE, including hard hats, gloves, goggles, and respirators, should be provided and maintained.
- Communication technologies: Advanced communication technologies, including wireless and satellite communication systems, should be implemented to improve communication between workers and management, especially in remote locations.
- Automation: Automation technologies, including driverless vehicles and autonomous drilling systems, can help reduce the need for workers to perform hazardous tasks, such as driving heavy equipment or operating drilling equipment.
- Sensors and monitoring systems: Advanced sensors and bio-monitoring systems can detect potential operation and human hazards, such as noise, fatigue, vibration, and gas leaks, and provide real-time information to workers and management.
- VR and AR: These technologies can be used to simulate dangerous mining conditions and train workers on how to respond to emergencies.
- Robotics: Robotics technologies can be used to perform hazardous tasks, such as inspecting mine shafts or loading explosives, reducing the need for workers to be exposed to dangerous situations.
- Predictive analytics: It can be used to anticipate potential safety hazards and mitigate them before they become serious risks.
- Continuous improvement: Finally, continuous improvement should be a top priority. The mining industry should continually evaluate and improve safety technologies, adapting to changing conditions and evolving risks.

A roadmap for mining safety technologies should prioritize risk assessment, education and training, PPE, communication, automation, sensors, monitoring, VR and AR, predictive analytics, robotics, and continuous improvement. By investing in these technologies, mining companies can reduce risks to the workers, the environment, and the surrounding community while continuing to provide important raw materials for a range of industry sectors.

Below is a table describing some of the technologies mentioned in this book with the potential impact to reduce unsafe conditions based on mining operational stages such as drilling, blasting, hauling, and processing.

Drilling	Blasting	Loading/Hauling/ Transporting	Crushing, Conveying, Smelting, Refining
• VR/AR training	• VR/AR training	• Interactive simulation model of human drivers' autonomous haulage systems	• VR based training program
• Directional drilling	• Remote/automated		• Wireless internet connection to improve miner safety
• Advisors located outside hazardous zones	• Ground control sensors	• Optimized miners' health by removing the operators from hazardous environments	• Machine learning to minimize human error
• Remote/automated	• Fume-less explosives		
• Through-ground communications tech	• Electro-mechanical blasting	• Remote/automated	• Remote/automated
• Rapid borehole drilling		• Vibration/jarring monitoring	• Biosensors

Figure 11.5 Safety technologies according to mining phases.

Drilling	Blasting	Loading/Hauling/ Transporting	Crushing, Conveying, Smelting, Refining
• Day rate model effect • Integrator for drilling systems automation	• Multi attribute decision making • Data envelopment analysis • Intelligent design system for mine blasting	• Autonomous loading: excavation, navigation, obstacle detection, obstacle avoidance • Trajectory studies of bucket motion • Computer modelling of excavation. • Computer simulations	• Theory of constrain analysis • Modular mining • Intellectualization of production process • Collaborative mining production • Automation

Drilling	Blasting	Loading/Hauling/ Transporting	Crushing, Conveying, Smelting, Refining
• Automated / remote drilling • Remote data centers • Smart alarms • AI systems • Embedded sensors • Machine learning • Robotics	• Robotics applications in bench blasting • AI and ANN, blast performance • AI and ANN, peak particle velocity analysis • Characterize blasting performance • Electro-mechanical fragmentation	• Advanced dedicated exhaustive communication network • AI based real-time dispatching • Automation • Remote operating centers • VR/AR applications	• AI, plant processes optimization (water, energy, emissions) • Automation • Embedded sensors • IoT, Big data, Analytics, Machine learning • Digital twins • Smart ore-sorting • Remote operating centers

Drilling	Blasting	Loading/Hauling/ Transporting	Crushing, Conveying, Smelting, Refining
• Wellsite monitoring systems. • Drilling dynamics diagnostic systems • MPD control systems. MWD rotary steerable Systems • Stick-slip surface control	• Mechanical fracturing • High-voltage-pulse discharge technology • Fume free explosives	• Zero emissions technology • Electrification • Hydrogen-electric • In-pit crushing and conveying systems	• Real-time implementation of micro-controllers to monitor emissions, noise • Power monitoring of metallurgical performance

Drilling	Blasting	Loading/Hauling/ Transporting	Crushing, Conveying, Smelting, Refining
• Reduction in fuel consumption • Reduce power consumption • Optimization of well footprint • Automation	• Efficient recovery of mineral by reducing the amount of waste • Predictable outcomes from precision blasting • Minimize use of explosives • Automation	• Automation • High level of driving and handling performance • Improve performance characteristics of the trucks • Decrease in fuel consumption • Increase tire lifetime • Increase production • Reduce fleet with less employed AHS	• Sensorial climate systems recycling • Efficient recovery processes • Eco-efficiency milling processes • Self contained recycled water systems • Energy use optimization • Footprint minimization • Renewable power sources

Figure 11.5 (Continued)

Simple-tech roadmap

Mining is a complex and highly technical process that involves the extraction of valuable minerals and resources from the earth. However, advancements in technology are simplifying mining processes and creating new opportunities for the industry. A simplification roadmap should prioritize efficient organizational policies, with an aggressive modernization strategy, promoting the use of simpler and more efficient technology such as electric mining systems. A scalable and modular strategy also needs to be considered to keep the overall operation adaptable to any contingencies.

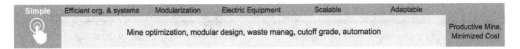

Figure 11.6 Simple-tech roadmap.

Here is a roadmap for mining that takes into consideration current and future simplification technologies.

Exploration: The first step in mining is to identify potential mineral deposits. Traditional exploration methods involve drilling and taking physical samples, which can be time-consuming and expensive. However, new technologies such as 3D seismic imaging, drone surveys, and satellite imagery can simplify the exploration process by providing more accurate and detailed data.

Site preparation: Once a potential mineral deposit has been identified, the site must be prepared for mining. This involves clearing vegetation, constructing access roads, and installing infrastructure such as power and water supply systems. Autonomous machinery and robots can simplify this process by reducing the need for human labor and improving safety.

Extraction: The actual process of mining involves extracting the mineral resource from the ground. Technologies such as automated drilling, blasting, and hauling can simplify this process by increasing efficiency and reducing the risk of accidents. In addition, advances in sensor technology and data analytics can help optimize the extraction process by providing real-time information on ore quality and quantity.

Processing: After the mineral resource has been extracted, it must be processed to extract the valuable minerals. Traditional processing methods involve grinding and chemical treatments, which can be expensive and environmentally damaging. However, new technologies such as bioreactors and hydrometallurgy can simplify the processing steps by using bacteria and other natural processes to extract minerals.

Reclamation: Once the mining process is complete, the site must be restored to its original condition. This involves filling in pits, re-vegetating the area, and monitoring the site for environmental impact. New technologies such as hydroseeding and drone monitoring can simplify the reclamation process by reducing the need for manual labor and providing more accurate data.

A simple-tech roadmap for mining involves embracing new technologies and approaches to simplify the mining process, increase efficiency, and reduce environmental impact. By leveraging these technologies, the mining industry can continue to extract valuable resources while reducing risk by being adaptable to any unpredicted economic, geological, or operating conditions, minimizing the financial risk.

Below is a table describing some of the technologies mentioned in this book with potential impact to optimize and simplify operational conditions based on mining operational phases such as drilling, blasting, hauling, and processing.

Drilling	Blasting	Loading/Hauling/Transporting	Crushing, Conveying, Smelting, Refining
• Day rate model effect • Integrator for drilling systems automation	• Multi attribute decision making • Data envelopment analysis • Intelligent design system for mine blasting	• Autonomous loading: Excavation, navigation, obstacle detection, obstacle avoidance • Trajectory studies of bucket motion • Computer modelling of excavation • Computer simulations	• Theory of constraint analysis • Modular mining • Intellectualization of production process • Collaborative mining production • Automation

Figure 11.7 Simplification technologies according to mining phases.

Smart-tech roadmap

A roadmap for smart mining that incorporates digital systems, machine learning, big data, AI, VR and AR, automation technologies, and other technologies might consider the following milestones considering current and future smart technologies.

Smart	Digitalization	Semi-automation	Automation	Smart Automation
	Data analytics, machine learning, sensors, smart automation, artificial ingeligence, virtual & augmented reality			Total Digital Twin Mine

Figure 11.8 Smart-tech roadmap.

Exploration and resource assessment

• Utilize AI and machine learning to process geospatial data to identify potential sites for mining.
• Use VR and AR technologies to visualize geological structures and analyze data to aid in exploration and resource assessment.

Extraction and mining operations

• Employ automation technologies to increase efficiency and reduce the risk of accidents in mining operations.
• Use drones equipped with sensors and cameras to monitor and inspect mining operations.
• Use AI to optimize mineral recovery rates and minimize waste.

Processing and refining

- Use automation to improve processing and refining efficiency.
- Utilize AI to predict and prevent equipment failures and optimize maintenance schedules.
- Use VR and AR to train workers and simulate scenarios for process improvement.

Environmental and social responsibility

- Use AI to monitor and minimize environmental impacts from mining operations.
- Use VR and AR to simulate and plan reclamation efforts and demonstrate progress to stakeholders.
- Utilize automation to reduce the risk of accidents and ensure worker safety.

Overall, the roadmap for mining that integrates digital, AI, VR, AR, and automation technologies has a significant technological advantage to improve operational efficiency, increase productivity, and promote growth. However, it is important to prioritize responsible and sustainable practices in the mining industry to ensure that these benefits will reduce environmental impacts and enhance worker safety and wellbeing.

Below is a table describing some of the technologies mentioned in this book with the potential impact to create a smart, digitally based operational condition based on mining operational phases such as drilling, blasting, hauling, and processing.

Drilling	Blasting	Loading/Hauling/ Transporting	Crushing, Conveying, Smelting, Refining
• Automated / remote drilling • Remote data centers • Smart alarms • AI systems • Embedded sensors • Machine learning • Robotics	• Robotics applications in bench blasting • AI and ANN, blast performance • AI and ANN, peak particle velocity analysis • Characterize blasting performance • Electro-mechanical fragmentation	• Advanced dedicated exhaustive communication network • AI based real-time dispatching • Automation • Remote operating centers • VR/AR applications	• AI, plant processes optimization (water, energy, emissions) • Automation • Embedded sensors • IoT, Big data, Analytics, Machine learning • Digital twins • Smart ore-sorting • Remote operating centers

Figure 11.9 Smart technologies according to mining phases.

Stealth-tech roadmap

The stealth technology roadmap in mining aims to reduce mining footprint by integrating advanced surface and underground mining methods, as well as by considering the use of inline underground mining, in-place underground ore processing, and advanced controlled in situ leaching, while optimizing performance and profitability.

Stealth	Underground processing mine	Inline, in-place underground extraction	In situ leaching	
○	Mine footprint minimization, underground mining, incremental open-pit, advanced caving			Invisible Mine

Figure 11.10 Stealth-tech roadmap.

Footprint minimization

To minimize the impact of mining on the environment, companies need to reduce their footprint through the use of advanced technologies, including the following:

- Remote sensing and geospatial analysis to identify the areas with the highest potential for mineral extraction while avoiding sensitive habitats and critical ecosystems.
- Drones and autonomous vehicles to conduct surveys and monitoring, reducing the need for physical infrastructure and human presence in the field.
- Tailings management systems that recycle water, reduce waste, and prevent the release of harmful chemicals into the environment.

Underground mining

Underground mining technologies can reduce the surface impact of mining, increase safety, and access deeper mineral deposits. Some of the technologies include the following:

- Automated drilling and blasting systems to increase productivity and reduce human exposure to hazards.
- Robotics and AI to improve efficiency and reduce downtime, improving safety and productivity.
- Advanced ventilation systems that improve air quality and reduce the risk of airborne hazards.

In situ processing and leaching

In situ processing technologies can reduce the need for traditional mining methods that require extensive excavation, transport, and processing of minerals. The following technologies can help reduce the environmental impact and lower the costs:

- In situ leaching, which involves pumping chemicals into the ground to dissolve minerals and recover them through a well.
- Electrowinning, which uses an electric current to separate minerals from ores, reducing the need for extensive excavation.
- Bioleaching, which uses microorganisms to extract metals from low-grade ores without the need for smelting.

Electric equipment

The use of electric equipment can reduce the carbon footprint of mining operations and lower costs through fuel savings. Some of the technologies include the following:

- Electric mining trucks and loaders, which can be powered by renewable energy sources such as solar or wind.
- Battery-powered equipment, which can operate underground without the need for ventilation or exhaust systems.
- Wireless charging systems, which can recharge equipment in transit, reducing downtime and increasing productivity.

Noise reduction

Mining operations can be a significant source of noise pollution, affecting both workers and nearby communities. Some of the technologies that can reduce noise levels include the following:

- Acoustic barriers and enclosures that can block noise sources and reduce the transmission of sound.
- Active noise control systems that use sensors and speakers to cancel out noise.
- Noise-reducing equipment such as mufflers and soundproof cabs for heavy machinery.

Automation

Automation technologies can improve efficiency, reduce labor costs, and improve safety in mining operations. Some of the technologies include the following:

- Autonomous haulage systems that use GPS and sensors to navigate and transport materials without human drivers.
- Robotic mining equipment that can operate in harsh environments and increase productivity.
- Remote control systems that allow operators to control equipment from a safe distance, reducing exposure to hazards.

A technology roadmap in mining should focus on minimizing the environmental footprint, improving safety, reducing costs, and increasing efficiency. A combination of innovative technologies and sustainable practices can help mining companies achieve these goals while meeting the increasing demand for minerals and metals.

Below is a table describing some of the technologies mentioned in this book with the potential impact to create a stealth, minimal-footprint operational condition based on mining operational phases such as drilling, blasting, hauling, and processing.

Drilling	Blasting	Loading/Hauling/ Transporting	Crushing, Conveying, Smelting, Refining
• Wellsite monitoring systems. • Drilling dynamics diagnostic systems • MPD control systems. MWD rotary steerable systems • Stick-slip surface control	• Mechanical fracturing • High-voltage-pulse discharge technology • Fume free explosives	• Zero emissions technology • Electrification • Hydrogen-electric • In-pit crushing and conveying systems	• Real-time implementation of micro-controllers to monitor emissions, noise • Power monitoring of metallurgical performance

Figure 11.11 Stealth technologies according to mining phases.

Sustainable-tech roadmap

A technology roadmap in mining that considers license to operate, human resource development, energy and water use minimization, recycling, and zero emissions would involve a series of steps to gradually transition the industry into more sustainable and environmentally friendly practices.

Here are some key milestones that could be included in such a roadmap.

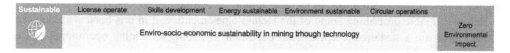

Figure 11.12 Sustainable-tech roadmap.

- Implementing advanced exploration techniques: Using advanced technologies such as remote sensing, drones, and geophysical surveys can help identify mineral deposits more accurately, and with less impact on the environment.
- Adopting best practices for land management: This would involve using responsible mining techniques, minimizing land disturbance, and working to restore ecosystems after mining operations are complete.
- Using renewable energy sources: Mining operations consume a significant amount of energy, and transitioning to renewable sources such as solar and wind power can help reduce the carbon footprint of mining activities.
- Developing zero-emission vehicles and equipment: Mining vehicles and equipment are a significant source of emissions and noise pollution. Developing and using electric, hybrid, or hydrogen-powered alternatives can help reduce this impact.
- Increasing the use of recycled materials: The mining industry could take steps to reuse and recycle materials, reducing the amount of virgin materials that need to be extracted from the earth.
- Investing in research and development: To achieve these goals, the mining industry would need to continue investing in research and development of new technologies and practices that can help reduce its environmental impact.
- Encouraging collaboration and transparency: Collaboration between mining companies, governments, and communities is important to ensure that mining operations are carried out responsibly and with minimal impact on the environment. This involves transparency in reporting and monitoring of environmental performance.

Drilling	Blasting	Loading/Hauling/ Transporting	Crushing, Conveying, Smelting, Refining
• Reduction in fuel consumption • Reduce power consumption • Optimization of well footprint • Automation	• Efficient recovery of mineral by reducing the amount of waste • Predictable outcomes from precision blasting • Minimize use of explosives • Automation	• Automation • High level of driving and handling performance • Improve performance characteristics of the trucks • Decrease in fuel consumption • Increase tire lifetime • Increase production • Reduce fleet with less employed AHS	• Sensorial climate systems recycling • Efficient recovery processes • Eco-efficiency milling processes • Self contained recycled water systems • Energy use optimization • Footprint minimization • Renewable power sources

Figure 11.13 Sustainable technologies according to mining phases.

By implementing these and other steps aimed to promote safety and health, the mining industry can work toward a more sustainable future, reducing its impact on the environment and improving the wellbeing of communities living near mining operations.

Below is a table describing some of the technologies mentioned in this book with potential impact to create a sustainable operational condition according to mining operational phases such as drilling, blasting, hauling, and processing.

Skills mine roadmap

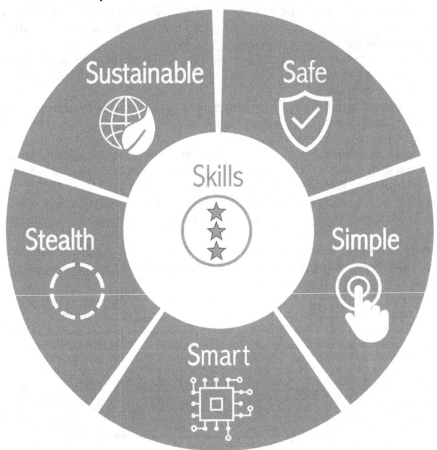

As discussed, the mining industry is undergoing a technology transformation, and companies are increasingly realizing the benefits of innovation. To stay competitive, mining companies need to adopt technology and innovation in line with their human resource policies. This is a short summary of a proposed skills-tech roadmap for a mining company that aims to adopt innovation and technology in its operations. A human skills development plan outlines the necessary steps to equip employees with the competencies and skills required to manage and implement new technologies and innovations effectively. The roadmap proposed includes the following key elements.

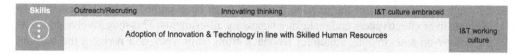

Figure 11.14 Skills-tech roadmap.

A Skills-tech roadmap may consist of four key stages: Implementing a new outreach and recruiting strategy, achieving an innovative thinking culture, embracing a new technology driven culture, and, finally, achieving an operation based on technology and innovation.

Implementing a new outreach and recruiting strategy

The first step in the technology skills-tech roadmap is to implement a new outreach and recruiting strategy. This strategy should focus on attracting talent with the right technology skills and mindset. The company should leverage social media and online job boards to reach a broader audience and make its job postings more visible.

The company should also partner with universities and vocational schools to identify and recruit young talent. It should provide internships, apprenticeships, and mentorship programs to give these individuals the opportunity to learn from the experienced professionals and gain hands-on experience.

Achieving an innovative thinking culture

The second stage of the roadmap is to achieve an innovative thinking culture. This involves creating an environment where employees are encouraged to think creatively and come up with new ideas. The company should invest in training programs that teach employees how to think outside the box, collaborate with others, and solve complex problems.

Training programs could consider the following:

1. Assess current skill levels: The first step in developing a human skills development plan is to assess the current skill levels of the employees. This can be achieved through interviews, surveys, and assessments to identify the areas where training and development are needed.
2. Identify training needs: Based on the assessment, the next step is to identify the specific training needs of the employees. This may include technical skills training, soft skills training, or a combination of both.
3. Develop training programs: Once the training needs have been identified, the organization must develop training programs that meet those needs. These programs may include classroom training, e-learning modules, or on-the-job training.
4. Assign trainers and mentors: Trainers and mentors play a crucial role in the development of employees' skills. They provide guidance and support throughout the training process and ensure that employees are developing the necessary skills to succeed in their roles.
5. Implement training programs: Once the training programs are developed, they must be implemented. This involves scheduling training sessions, providing employees with access to the necessary resources, and monitoring their progress.

6. Evaluate and adjust: After the training programs have been implemented, it is essential to evaluate their effectiveness and make any necessary adjustments. This may involve conducting follow-up assessments, soliciting feedback from employees, and revising the training programs based on the results.

7. Maintain ongoing training: Finally, to ensure that employees continue to develop their skills, the organization must maintain ongoing training programs. This may involve providing refresher courses, introducing new technologies, and offering opportunities for employees to attend industry conferences and workshops.

The company should also create a platform where employees can share their ideas and get feedback from others. This platform should be open to all employees and should encourage a culture of collaboration and innovation.

Embracing a new innovation and technology culture

The third stage of the roadmap is to embrace a new innovation and technology culture. This involves introducing new technologies and tools that enable employees to work more efficiently and effectively. The company should invest in technology that improves communication and collaboration among employees, such as cloud-based tools, video conferencing, and collaboration software.

The company should also invest in technologies that improve safety and reduce environmental impact, such as autonomous vehicles and sensors that monitor air and water quality as discussed previously in this book. The company should provide training programs to teach employees how to use these technologies effectively and safely.

Achieving an operation based on technology and innovation

The final stage of the roadmap is to achieve an operation based on technology and innovation. This involves leveraging technology to improve every aspect of the company's operations, from exploration and development to production and logistics. The company should use data analytics and AI to optimize its operations and make data-driven decisions.

The company should also invest in automation to reduce costs and improve efficiency. This could include using drones to survey mine sites, autonomous trucks to transport materials, and robots to perform maintenance tasks.

A technology skills-tech roadmap is essential for mining companies that want to adopt innovation and technology in their operations. The roadmap should focus on implementing a new outreach and recruiting strategy, achieving an innovative thinking culture, embracing a new innovation and technology culture, and finally achieving an operation based on technology and innovation. By following this roadmap, mining companies can stay competitive in a rapidly evolving industry and attract and retain top talent.

By incorporating human skills development into a technology roadmap, organizations can ensure that their workforce has the necessary skills and knowledge to effectively manage new technologies and innovations. This, in turn, can help the organization stay competitive, increase productivity, and achieve its strategic goals.

Further reading

Bascetin, A., S. Tuylu, A. Nieto (2011), Influence of the ore block model estimation on the determination of the mining cutoff grade policy for sustainable mine production. *Environmental Earth Sciences*, Springer, Berlin, vol. 65, no. 5, pp. 1409–1418.

Chitombo, G. (2018), Cave mining 2040: Advancing cave mining to meet future needs. *mining3*, Austmine Limited, Monday, 2 July, Categories: Articles & Editorials.

CRC-Mining (2018), In place mining – A transformational shift in metal extraction. Web: www.mining3.com/place-mining-transformational-shift-metal-extraction/

FLS Services (2022), Typical mining and mineral process flowsheet, Public Domain. Web: www.flsmidth.com/en-gb/solutions/optimise-plant-performance-through-remote-continuous-monitoring

Galantucci, L., M. Guerra, M. Dassisti, F. Lavecchia (2019), Additive manufacturing: New trends in the 4th industrial revolution. *Proceedings of the 4th International Conference on the Industry 4.0 Model for Advanced Manufacturing.* https://doi.org/10.1007/978-3-030-18180-2_12.

National Academy of Engineering (2018), Grand challenges of engineering report, National academy of sciences on behalf of the national academy of engineering. Web: www.engineeringchallenges.org/

Nieto, A. (2010), Key deposit indicators (KDI) and key mining method indicators (KMI) in underground mining method selection. *Transactions of the Society for Mining, Metallurgy and Engineering*, Inc., Littleton, CO, vol. 328, pp. 381–396.

Nieto, A. (2011), Selection process for underground soft-rock. In *SME Mining Engineering Handbook*, 3rd ed. Society for Mining, Metallurgy and Exploration. ISBN 1613440413, 9781613440414

Nieto, A. (2019a), The mine of the future: The 5S innovation model for the minerals industry. *GeoResources Journal*, vol. 3, no. 2019, pp. 35–39.

Nieto, A. (2019b), *Research Notes, School of Mining Engineering*. Wits University.

Nieto, A., K. Dagdelen (2001), Improving safety of off highway trucks through *(APCOM*) 2001*, Beijing, CH, pp. 757–762.

Nieto, A., M. Lannuzzi (2012), Supply and demand geo-economic analysis of mineral resources of rare earth elements in the United States. *SME, Mining Magazine*, vol. 64, no. 4, pp. 74–82.

Nieto, A., J. Medina (2020), Development of a socioeconomic strategic risk index as an aid for the feasibility assessment of mining projects and operations. *The Journal of the Southern African Institute of Mining and Metallurgy*, vol. 120, no. 7. http://doi.org/10.17159/2411-9717/711/2020.

Schatz, R., A. Nieto, C. Dogruoz, S. Lvov (2015), Using modern battery systems in light duty vehicles. *International Journal of Mining, Reclamation & Environment*, vol. 29, no. 4, pp. 243–265. https://doi.org/10.1080/17480930.2013.866797

Schatz, R., A. Nieto, S. Lvov (2017), Long-term economic sensitivity analysis of light duty underground mining vehicles by power source. *International Journal of Mining Science and Technology*, Elsevier, vol. 25, no. 3, pp. 567–571.

Schumpeter, J.A. (1954), *History of Economic Analysis*. Psychology Press.

Sinha, S., S. Chakraborty, D. Shome (2019), Mining footprint: A spatial indicator of environmental quality – a case study of a manganese mine in Bhandara district. *Maharashtra*, vol. 12, p. 96. https://doi.org/10.1007/s12517-019-4260-0

Stamm, M.L., T. Neitzert, P.K. Singh (2019), *TQM, TPM, TOC, Lean and Six Sigma – Evolution of Manufacturing Methodologies Under the Paradigm Shift from Taylorism/Fordism to Toyotism?* https://core.ac.uk/download/pdf/56363032.pdf

Xenaki, A., A. Nieto, E. Laporte, J. Castillo, O. Velasquez (2023), Application of a tech innovation model for the mine of the future: Bridging the gap between theory and practice. *2023 SME Annual Conference & Expo, Society of Mining, Metallurgy & Exploration*, Denver, Colorado, February 26–March 1.

Index

Printed in the United States
by Baker & Taylor Publisher Services